ELECTRIC
and
HYBRID
VEHICLES
Design Fundamentals

ELECTRIC and HYBRID VEHICLES

Design Fundamentals

Iqbal Husain

CRC PRESS

Boca Raton   London   New York   Washington, D.C.

Library of Congress Cataloging-in-Publication Data

Husain, Iqbal, 1964-
    Electric and hybrid vehicles : design fundamentals / by Iqbal Husain.
        p. cm.
    Includes bibliographical references and index.
    ISBN 0-8493-1466-6 (alk. paper)
        1. Electric vehicles. 2. Hybrid electric vehicles. I. Title.

TL220 .H87 2003
629.22'93--dc21   2002041120
                                                                   CIP

Visit the CRC Press Web site at www.crcpress.com

© 2003 by CRC Press LLC

No claim to original U.S. Government works
International Standard Book Number 0-8493-1466-6
Library of Congress Card Number 2002041120
Printed in the United States of America   5 6 7 8 9 0
Printed on acid-free paper

Preface

The book presents a comprehensive systems-level perspective of electric and hybrid electric vehicles, with emphasis on technical details, mathematical relationships, and basic design guidelines. The electric vehicle is an excellent example of an electro-mechanical and electrochemical system that is technically challenging as well as highly intriguing to engineering students. With a good balance between technical details, design equations, numerical examples, and case studies, the subject matter presents an ideal platform for educating today's engineers with a systems-level perspective — a concept that served as the primary motivation to develop this textbook on electric and hybrid vehicles.

Automobiles are an integral part of our everyday lives. Yet, conventional automobiles are the major cause of urban pollution in the 21st century. The world will eventually encounter an acute energy crisis if we do not focus on alternative energy sources and transportation modes. Current environmental concerns are driving the international community toward developing low-emission (hybrid electric) and zero-emission (electric) vehicles to replace conventional internal combustion engine vehicles. The subject of electric and hybrid vehicles is becoming increasingly important, with intense drive from the government, environmental activists, and associated industries to advance the technology. Several auto industries have already started marketing electric and hybrid electric vehicles. Furthermore, the next generation of conventional automobiles will experience a gradual replacement of the hydraulically driven actuators by electrically driven actuators. The trend clearly suggests that there is a need to adequately educate the engineers of today and tomorrow with the technical details of electric and hybrid vehicles and the electrical units used within an automobile. While there are ample books on electric and hybrid vehicles available, providing narrative descriptions of the components of vehicles, and numerous technical papers published with research results, none covers the technical aspects and mathematical relationships in a comprehensive way to educate a junior- or senior-level or a beginning graduate-level engineering student.

This book will serve to educate students on aspects of electric vehicles, which will generate interest to support the development and use of electric vehicles. The book will also serve as a reference for a working engineer dealing with design and improvement of electric and hybrid vehicles. Discussion on most topics has been limited to fundamentals only in the book, considering the wide spectrum of technical aspects related to an electric and hybrid vehicle system. Appropriate references are given to direct the readers toward details on topics for further reading. The intent of the book is not to present the wide spectrum of the state of the art in electric and hybrid electric vehicles, but rather to prepare the student with the necessary background to evaluate the technology.

The book, starting with a historical background on electric vehicles, will describe the system components, the laws of physics that govern vehicle motion, the mathematical relationships within a component and between components, the energy sources, and the design of components to meet the specifications for the complete vehicle system. After the introduction of the systems concept in Chapter 1, Chapter 2 focuses on the laws of physics to define the force characteristics of ground vehicles. The design guidelines for the power and energy requirements based on design specifications are established in this chapter.

The flow of the book shifts from mechanical to chemical concepts, when energy sources are introduced in Chapter 3, and the topic is continued in Chapter 4, with alternatives to battery power. The two major contenders for energy sources in road vehicles are batteries and fuel cells, which are described in detail, while other types of energy sources are mentioned briefly.

Chapters 5 through 8 are mostly electrical, where electric motors for propulsion and power electronic drives for the motors are presented. The DC machines and AC induction machines suitable for propulsion are discussed in Chapter 5, while the permanent magnet and switched reluctance machines are presented in Chapter 6. Chapters 7 and 8 are dedicated to the power-electronics-based motor drives for electric propulsion units. Vehicle system control fundamentals are also addressed in these two chapters.

Mechanical and electrical concepts merge in Chapters 9 and 10. Drivetrain components, including the transmission for electric vehicles, are presented in Chapter 9, while Chapter 10 discusses the drivetrain and the design basics of hybrid electric vehicles.

This book is intended to be used as a textbook for an undergraduate or beginning graduate-level course on electric and hybrid electric vehicles. The ten chapters of the book can be comfortably covered in a three-credit, one-semester or a four-credit, one-quarter course. Although the materials in this book are biased toward the electrical units, it is still multidisciplinary enough to teach electrical, mechanical, and chemical engineers all in one course, utilizing the systems approach. In that case, parts of the electrical details appearing in Chapters 5 though 8 should be skipped. This type of course will certainly mimic the real situation existing in many industries, where multidisciplinary engineers work together to devise a system and develop a product. The equations developed can be utilized to develop a system-level modeling and simulation tool for electric and hybrid electric vehicles on a suitable platform, such as MATLAB/SIMULINK. The book has several worked-out problems and many exercises that are suitable to convey the concept to students through numerical examples.

Author

Dr. Iqbal Husain is an Associate Professor in the Department of Electrical and Computer Engineering at the University of Akron, Akron, Ohio, where he is engaged in teaching and research. After earning his Ph.D. degree in Electrical Engineering from Texas A&M University, College Station, in 1993, Dr. Husain worked as a lecturer at Texas A&M University and as a consulting engineer for Delco Chassis at Dayton, Ohio, prior to joining the University of Akron in 1994. He worked as a summer researcher for Wright Patterson AFB Laboratories in 1996 and 1997. More recently, he taught at Oregon State University as a short-term visiting faculty member. His research interests are in the areas of control and modeling of electrical drives, design of electric machines, and development of power conditioning circuits. He has worked extensively in the development of switched reluctance motor drives, including sensorless controllers. He also worked as a consultant for Delphi Automotive Systems, Goodyear Tire and Rubber Industry, ITT Automotive, Delphi Chassis, Graphic Enterprises, and Hy-Tech Inc.

Dr. Husain received the 2000 IEEE Third Millenium Medal, the 1998 IEEE-IAS Outstanding Young Member award, and the NSF CAREER Award in 1997. He is also the recipient of three IEEE Industry Applications Society prize paper awards.

Acknowledgments

I would like to express my sincere gratitude to all those who helped me devotedly to complete the work. I would like to thank my former and current graduate students, John Bates, Liu Tong, Nazmul Anwar, Shahidul Islam, Afjal Hossain, Faizul Momen, Virginie Raulin, Mihaela Radu, Ahmed Khalil, and Jin Wang, who helped me tremendously with problems, figures, and materials. I offer my gratitude to Dr. Don Zinger, who first offered a course on electric vehicles at the University of Akron, Akron, Ohio, and created an opportunity for me to prepare textbook materials on the topic.

I am extremely thankful to Dr. Robert Pasch of Oregon State University and Dr. Richard Gross of the University of Akron, both from Mechanical Engineering Departments, who educated and helped me in writing about the mechanical-related topics. I would also like to thank the reviewers who provided extremely useful suggestions that helped enhance the quality of the book. The reviewers included Prof. Alan K. Wallace and Prof. Annette von Jouanne, Department of Electrical and Computer Engineering, Oregon State University; Prof. M. Ehsani, Department of Electrical Engineering, Texas A&M University; Prof. Longya Xu, Department of Electrical and Computer Engineering, Ohio State University; Prof. Pragassen Pillay, Department of Electrical and Computer Engineering, Clarkson University; Dr. Khwaja M. Rahman, General Motors ATV; and Dr. Alexander Yokochi, Department of Chemistry, Oregon State University.

I thank the staff of CRC Press LLC, especially Nora Konopka and Helena Redshaw, whose guidance was invaluable in preparing my first textbook manuscript. Finally, my sincere apologies and heartfelt gratitude to my wife, Salina, and my children Inan and Imon, who patiently stood by me with grave understanding and continuous support while I was preoccupied with the project.

Iqbal Husain
Akron, Ohio

Table of Contents

1 Introduction to Electric Vehicles

Environmental as well as economical issues provide a compelling impetus to develop clean, efficient, and sustainable vehicles for urban transportation. Automobiles constitute an integral part of our everyday life, yet the exhaust emissions of conventional internal combustion (IC) engine vehicles are to blame for the major source of urban pollution that causes the greenhouse effect leading to global warming.[1] The dependence on oil as the sole source of energy for passenger vehicles has economical and political implications, and the crisis will inevitably become acute as the oil reserve of the world diminishes. The number of automobiles on our planet doubled to about a billion or so in the last 10 years. The increasing number of automobiles being introduced on the road every year is only adding to the pollution problem. There is also an economic factor inherent in the poor energy conversion efficiency of combustion engines. Although the number for alternative electric vehicles is not significantly higher when efficiency is evaluated on the basis of conversion from crude oil to traction effort at the wheels, it makes a difference. Emission due to power generation at localized plants is much easier to regulate than that emanating from IC engine vehicles (ICEV) that are individually maintained and scattered. People dwelling in cities are not exposed to power plant related emissions, because these are mostly located outside urban areas. Electric vehicles (EV) enabled by high-efficiency electric motors and controllers and powered by alternative energy sources provide the means for a clean, efficient, and environmentally friendly urban transportation system. Electric vehicles have no emission, having the potential to curb the pollution problem in an efficient way. Consequently, EVs are the only zero-emission vehicles possible.

Electric vehicles paved their way into public use as early as the middle of the 19th century, even before the introduction of gasoline-powered vehicles.[2] In the year 1900, 4200 automobiles were sold, out of which 40% were steam powered, 38% were electric powered, and 22% were gasoline powered. However, the invention of the starter motor, improvements in mass production technology of gas-powered vehicles, and inconvenience in battery charging led to the disappearance of the EV in the early 1900s. However, environmental issues and the unpleasant dependence on oil led to the resurgence of interest in EVs in the 1960s. Growth in the enabling technologies added to environmental and economic concerns over the next several decades, increasing the demand for investing in research and development for EVs. Interest and research in EVs soared in the 1990s, with the major automobile manufacturers embarking on plans for introducing their own electric or hybrid electric

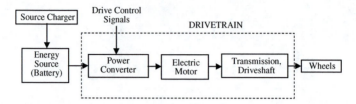

FIGURE 1.1 Top-level perspective of an EV system.

vehicles. The trend increases today, with EVs serving as zero-emission vehicles, and hybrid electric vehicles already filling in for ultralow-emission vehicles.

1.1 EV SYSTEM

An EV has the following two features:

1. The energy source is portable and chemical or electromechanical in nature.
2. Traction effort is supplied only by an electric motor.

Figure 1.1 shows an EV system driven by a portable energy source. The electromechanical energy conversion linkage system between the vehicle energy source and the wheels is the drivetrain of the vehicle. The drivetrain has electrical as well as mechanical components.

1.1.1 COMPONENTS OF AN EV

The primary components of an EV system are the motor, controller, power source, and transmission. The detailed structure of an EV system and the interaction among its various components are shown in Figure 1.2. Figure 1.2 also shows the choices available for each of the subsystem level components. Electrochemical batteries have been the traditional source of energy in EVs. Lead–acid batteries have been the primary choice, because of their well-developed technology and lower cost, although promising new battery technologies are being tested in many prototype vehicles. The batteries need a charger to restore the stored energy level once its available energy is near depletion due to usage. Alternative energy sources are also being developed for zero-emission vehicles. The limited range problem of battery-driven EVs prompted the search for alternative energy sources, such as fuel cells and flywheels. Prototypes have been developed with fuel cells, while production vehicles will emerge in the near future.

The majority of electric vehicles developed so far are based on DC machines, induction machines, or permanent magnet machines. The disadvantages of DC machines pushed EV developers to look into various types of AC machines. The maintenance-free, low-cost induction machines became an attractive alternative to many developers. However, high-speed operation of induction machines is only possible with a penalty in size and weight. Excellent performance together with

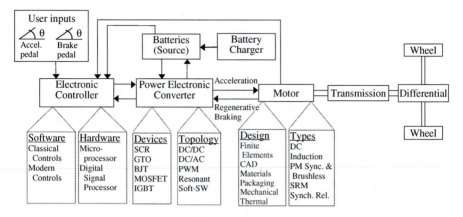

FIGURE 1.2 Major electrical components and choices for an EV system.

high-power density features of permanent magnet machines make them an attractive solution for EV applications, although the cost of permanent magnets can become prohibitive. High-power density and a potentially low production cost of switched reluctance machines make them ideally suited for EV applications. However, the acoustic noise problem has so far been a deterrent for the use of switched reluctance machines in EVs. The electric motor design includes not only electromagnetic aspects of the machine but also thermal and mechanical considerations. The motor design tasks of today are supported by finite element studies and various computer-aided design tools, making the design process highly efficient.

The electric motor is driven by a power-electronics-based power-processing unit that converts the fixed DC voltage available from the source into a variable voltage, variable frequency source controlled to maintain the desired operating point of the vehicle. The power electronics circuit comprised of power semiconductor devices saw tremendous development over the past 3 decades. The enabling technology of power electronics is a key driving force in developing efficient and high-performance power-train units for EVs. High-power devices in compact packaging are available today, enabling the development of lightweight and efficient power-processing units known as power electronic motor drives. Advances in power solid state devices and very large-scale integration (VLSI) technology are responsible for the development of efficient and compact power electronics circuits. The developments in high-speed digital signal processors or microprocessors enable complex control algorithm implementation with a high degree of accuracy. The controller includes algorithms for the motor drive in the inner loop as well as system-level control in the outer loop.

1.2 EV HISTORY

The history of EVs is interesting. It includes the insurgence of EVs following the discovery of electricity and the means of electromechanical energy conversion and later being overtaken by gasoline-powered vehicles. People digressed from the environmentally friendly mode of transportation due to lack of technology in the early years, but they are again focused on the correct track today.

1.2.1 THE EARLY YEARS

Prior to the 1830s, the means of transportation was only through steam power, because the laws of electromagnetic induction, and consequently, electric motors and generators, were yet to be discovered. Faraday demonstrated the principle of the electric motor as early as in 1820 through a wire rod carrying electric current and a magnet, but in 1831 he discovered the laws of electromagnetic induction that enabled the development and demonstration of the electric motors and generators essential for electric transportation. The history of EVs in those early years up to its peak period in the early 1900s is summarized below:

- Pre-1830 — Steam-powered transportation
- 1831 — Faraday's law, and shortly thereafter, invention of DC motor
- 1834 — Nonrechargeable battery-powered electric car used on a short track
- 1851 — Nonrechargeable 19 mph electric car
- 1859 — Development of lead storage battery
- 1874 — Battery-powered carriage
- Early 1870s — Electricity produced by dynamo-generators
- 1885 — Gasoline-powered tricycle car
- 1900 — 4200 automobiles sold:
  - 40% steam powered
  - 38% electric powered
  - 22% gasoline powered

The specifications of some of the early EVs are given below:

- 1897 — French Krieger Co. EV: weight, 2230 lb; top speed, 15 mph; range, 50 mi/charge
- 1900 — French B.G.S. Co. EV: top speed, 40 mph; range, 100 mi/charge
- 1912 — 34,000 EVs registered; EVs outnumber gas-powered vehicles 2-to-1
- 1915 — Woods EV: top speed, 40 mph; range, 100 mi/charge
- 1915 — Lansden EV: weight, 2460 lb, top speed, 93 mi/charge, capacity, 1 ton payload
- 1920s — EVs disappear, and ICEVs become predominant

The factors that led to the disappearance of EV after its short period of success were as follows:

1. Invention of starter motor in 1911 made gas vehicles easier to start.
2. Improvements in mass production of Henry T (gas-powered car) vehicles sold for $260 in 1925, compared to $850 in 1909. EVs were more expensive.

3. Rural areas had limited access to electricity to charge batteries, whereas gasoline could be sold in those areas.

1.2.2   1960s

Electric vehicles started to resurge in the 1960s, primarily due to environmental hazards being caused by the emissions of ICEVs. The major ICEV manufacturers, General Motors (GM) and Ford, became involved in EV research and development. General Motors started a $15 million program that culminated in the vehicles called Electrovair and Electrovan. The components and specifications of two Electrovair vehicles (Electrovair I (1964) and Electrovair II (1966) by GM) are given below.

Systems and characteristics:
      Motor — three-phase induction motor, 115 hp, 13,000 rev/m
      Battery — silver-zinc (Ag-Zn), 512 V, 680 lb
      Motor drive — DC-to-AC inverter using a silicon-controlled rectifier
        (SCR)
      Top speed — 80 mi/h
      Range — 40 to 80 miles
      Acceleration — 0–60 mi/h in 15.6 s
      Vehicle weight — 3400 lb

The Electrovair utilized the Chevy Corvair body and chassis. Among the positive features was the acceleration performance that was comparable to the ICEV Corvair. The major disadvantage of the vehicle was the silver-zinc (Ag-Zn) battery pack that was too expensive and heavy, with a short cycle life and a long recharge time.

An additional factor in the 1960s that provided the impetus for EV development included "The Great Electric Car Race" cross-country competition (3300 miles) between an EV from Caltech and an EV from MIT in August 1968. The race generated great public interest in EVs and provided an extensive road test of the EV technology. However, technology of the 1960s was not mature enough to produce a commercially viable EV.

1.2.3   1970s

The scenario turned in favor of EVs in the early 1970s, as gasoline prices increased dramatically due to an energy crisis. The Arab oil embargo of 1973 increased demands for alternate energy sources, which led to immense interest in EVs. It became highly desirable to be less dependent on foreign oil as a nation. In 1975, 352 electric vans were delivered to the U.S. Postal Service for testing. In 1976, Congress enacted Public Law 94–413, the *Electric and Hybrid Vehicle Research, Development and Demonstration Act of 1976*. This act authorized a federal program to promote electric and hybrid vehicle technologies and to demonstrate the commercial feasibility of EVs. The Department of Energy (DOE) standardized EV performance, which is summarized in Table 1.1.

TABLE 1.1
EV Performance Standardization of 1976

Category	Personal Use	Commercial Use
Acceleration from 0 to 50 km/h	<15 s	<15 s
Gradability at 25 km/h	10%	10%
Gradability at 20 km/h	20%	20%
Forward speed for 5 min	80 km/h	70 km/h
Range:		
Electric	50 km, C cycle	50 km, B cycle
Hybrid	200 km, C cycle	200 km, B cycle
Nonelectrical energy consumption in hybrid vehicles (consumption of nonelectrical energy must be less than 75% of the total energy consumed)	<1.3 MJ/km	<9.8 MJ/km
Recharge time from 80% discharge	<10 h	<10 h

The case study of a GM EV of the 1970s is as follows:

System and characteristics:
    Motor — separately excited DC, 34 hp, 2400 rev/m
    Battery pack — Ni-Zn, 120 V, 735 lb
    Auxiliary battery — Ni-Zn, 14 V
    Motor drive — armature DC chopper using SCRs; field DC chopper using
        bipolar junction transistors (BJTs)
    Top speed — 60 mi/h
    Range — 60–80 miles
    Acceleration — 0–55 mi/h in 27 s

The vehicle utilized a modified Chevy Chevette chassis and body. This EV was used mainly as a test bed for Ni-Zn batteries. Over 35,500 miles of on-road testing proved that this EV was sufficiently road worthy.

1.2.4 1980s AND 1990s

In the 1980s and the 1990s, there were tremendous developments of high-power, high-frequency semiconductor switches, along with the microprocessor revolution, which led to improved power converter design to drive the electric motors efficiently. Also in this period, factors contributed to the development of magnetic bearings used in flywheel energy storage systems, although these are not utilized in mainstream EV development projects.

In the last 2 decades, legislative mandates pushed the cause for zero-emission vehicles. Legislation passed by the California Air Resources Board in 1990 stated that by 1998 2% of vehicles should be zero-emission vehicles (ZEV) for each automotive company selling more than 35,000 vehicles. The percentages were to increase to 5% by 2001 and to 10% by 2003. The legislation provided a tremendous impetus to develop EVs by the major automotive manufacturers. The legislation was relaxed somewhat later due to practical limitations and the inability of the manufacturers to meet the 1998 and 2001 requirements. The mandate now stands that 4% of all vehicles sold should be ZEV by 2003, and an additional 6% of the sales must be made up of ZEVs and partial ZEVs, which would require GM to sell about 14,000 EVs in California.

Motivated by the pollution concern and potential energy crisis, government agencies, federal laboratories, and the major automotive manufacturers launched a number of initiatives to push for ZEVs. The partnership for next-generation vehicles (PNGV) is such an initiative (established in 1993), which is a partnership of federal laboratories and automotive industries to promote and develop electric and hybrid electric vehicles. The most recent initiative by the DOE and the automotive industries is the Freedom CAR initiative.

The trends in EV developments in recent years can be attributed to the following:

- High level of activity exists at the major automotive manufacturers.
- New independent manufacturers bring vigor.
- New prototypes are even better.
- High levels of activity overseas exist.
- There are high levels of hybrid vehicle activity.
- A boom in individual ICEV to EV conversions is ongoing.
- The fuel cell shows great promise in solving the battery range problem.

The case studies of two GM EVs of the 1990s are given below:

1. GM Impact 3 (1993 completed):
   a. Based on 1990 Impact displayed at the Los Angeles auto show
   b. Two-passenger, two-door coupe, street legal and safe
   c. Initially, 12 built for testing; 50 built by 1995 to be evaluated by 1000 potential customers
   d. System and characteristics:
      i. Motor — one, three-phase induction motor; 137 hp; 12,000 rev/m
      ii. Battery pack — lead-acid (26), 12 V batteries connected in series (312 V), 869 lb
      iii. Motor drive — DC-to-AC inverter using insulated gate bipolar transistors (IGBTs)
      iv. Top speed — 75 mph
      v. Range — 90 miles on highway
      vi. Acceleration — 0 to 60 miles in 8.5 s
      vii. Vehicle weight — 2900 lb
   e. This vehicle was used as a test bed for mass production of EVs.
2. Saturn EV1
   a. Commercially available electric vehicle made by GM in 1995.
   b. Leased in California and Arizona for a total cost of about $30,000.

    c. System and characteristics:
       i. Motor — one, three-phase induction motor
       ii. Battery pack — lead-acid batteries
       iii. Motor drive — DC-to-AC inverter using IGBTs
       iv. Top speed — 75 mph
       v. Range — 90 miles on highway, 70 miles in city
       vi. Acceleration — 0 to 60 mi in 8.5 s
    d. Power consumption:
       i. 30 kW-h/100 mi in city, 25 kW-h/100 mi on highway
    e. This vehicle was also used as a test bed for mass production of EVs.

1.2.5 RECENT EVs AND HEVs

All of the major automotive manufacturers have production EVs, many of which are available for sale or lease to the general public. The status of these vehicle programs changes rapidly, with manufacturers suspending production frequently due to the small existing market demand of such vehicles. Examples of production EVs which are or until recently have been available are GM EV1, Ford Think City, Toyota RAV4, Nissan Hypermini, and Peugeot 106 Electric. There are also many prototype and experimental EVs being developed by the major automotive manufacturers. Most of these vehicles use AC induction motors or PM synchronous motors. Also, interestingly, almost all of these vehicles use battery technology other than the lead-acid battery pack. The list of EVs in production and under development is extensive, and readers are referred to the literature[3,4] for the details of many of these vehicles.

The manufacturers of EVs in the 1990s realized that their significant research and development efforts on ZEV technologies were hindered by unsuitable battery technologies. A number of auto industries started developing hybrid electric vehicles (HEVs) to overcome the battery and range problem of pure electric vehicles. The Japanese auto industries lead this trend with Toyota, Honda, and Nissan already marketing their Prius, Insight, and Tino model hybrids. The hybrid vehicles use an electric motor and an internal combustion engine and, thus, do not solve the pollution problem, although it does mitigate it. It is perceived by many that the hybrids, with their multiple propulsion units and control complexities, are not economically viable in the long run, although currently a number of commercial, prototype, and experimental hybrid vehicle models are available from almost all of the major automotive industries around the world. Toyota, Honda, and Nissan are marketing the hybrid vehicles well below the production cost, with significant subsidy and incentive from the government. However, the cost of HEVs and EVs are expected to be high until production volume increases significantly.

Fuel cell electric vehicles (FCEV) can be a viable alternative to battery electric vehicles, serving as zero-emission vehicles without the range problem. Toyota is leading the way with FCEV, announcing the availability of its FCEV in 2003. The Toyota FCEV is based on the Toyota RAV4 model.

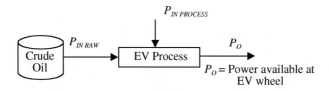

FIGURE 1.3 EV process from crude oil to power at the wheels.

1.3 EV ADVANTAGES

The relative advantages and disadvantages of an EV over an ICEV can be better appreciated from a comparison of the two on the bases of efficiency, pollution, cost, and dependence on oil. The comparison must be executed with care, ensuring fairness to both systems.

1.3.1 EFFICIENCY COMPARISON

To evaluate the efficiencies of EV and ICEV on level ground, the complete process in both systems starting from crude oil to power available at the wheels must be considered. The EV process starts not at the vehicles, but at the source of raw power whose conversion efficiency must be considered to calculate the overall efficiency of electric vehicles. The power input P_{IN} to the EV comes from two sources — the stored power source and the applied power source. Stored power is available during the process from an energy storage device. The power delivered by a battery through electrochemical reaction on demand or the power extracted from a piece of coal by burning it are examples of stored power. Applied power is obtained indirectly from raw materials. Electricity generated from crude oil and delivered to an electric car for battery charging is an example of applied power. Applied power is labeled as $P_{IN\,RAW}$, while stored power is designated as $P_{IN\,PROCESS}$ in Figure 1.3. Therefore, we have the following:

$$P_{IN} = P_{IN\,PROCESS} + P_{IN\,RAW}$$

The complete EV process can be broken down into its constituent stages involving a chain of events responsible for power generation, transmission, and usage, as shown in Figure 1.4. Raw power from the applied source is fed to the system only at the first stage, although stored power can be added in each stage. Each stage has its efficiency based on total input to that stage and output delivered to the following stage. For example, the efficiency of the first stage based on the input and output shown in Figure 1.4 is

$$\eta_1 = \frac{P_1}{P_{IN\,RAW} + P_{IN\,PROCESS1}}$$

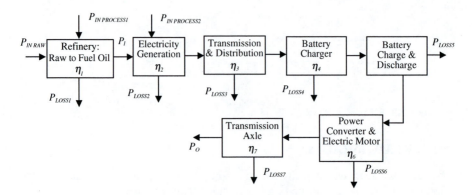

FIGURE 1.4 The complete EV process broken into stages.

The efficiency of each stage must be calculated from input–output power considerations, although the efficiency may vary widely, depending on the technology being used. Finally, overall efficiency can be calculated by multiplying the efficiencies of the individual stages. The overall efficiency of the EV system shown in Figure 1.4 is

$$\eta_{EV} = \frac{P_0}{P_{IN}} = \frac{P_0}{P_0 + \sum_{i=1}^{7} P_{LOSSi}} = \eta_1\eta_2\eta_3\eta_4\eta_5\eta_6\eta_7$$

The overall ICEV process is shown in Figure 1.5, while the process details are illustrated in Figure 1.6. Starting from the conversion of crude oil to fuel oil in the refinery, the ICEV process includes the transmission of fuel oil from refinery to gas stations, power conversion in the internal combustion engine of the vehicle, and power transfer from the engine to the wheels through the transmission before it is available at the wheels. The efficiency of the ICEV process is the product of the efficiencies of the individual stages indicated in Figure 1.6 and is given by

$$\eta_{ICEV} = \eta_1\eta_2\eta_3\eta_4$$

A sample comparison of EV and ICEV process efficiencies based on the diagrams of Figure 1.4 and 1.6 is given in Table 1.2. Representative numbers have been used for the energy conversion stages in each process to convey a general idea of the efficiencies of the two systems. From Table 1.2, it can be claimed that the overall efficiency of an EV is comparable to the overall efficiency of an ICEV.

1.3.2 POLLUTION COMPARISON

Transportation accounts for one third of all energy usage, making it the leading cause of environmental pollution through carbon emissions.[5] The DOE projected that if 10% of automobiles nationwide were zero-emission vehicles, regulated air

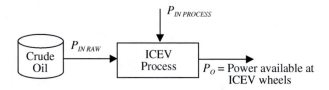

FIGURE 1.5 ICEV process from crude oil to power at the wheels.

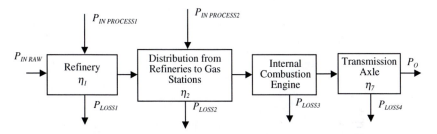

FIGURE 1.6 The complete ICEV process broken into stages.

pollutants would be cut by 1,000,000 tons per year, and 60,000,000 tons of greenhouse carbon dioxide gas would be eliminated. With 100% electrification, i.e., every ICEV replaced by an EV, the following was claimed:

- Carbon dioxide in air, which is linked to global warming, would be cut in half.
- Nitrogen oxides (a greenhouse gas causing global warming) would be cut slightly, depending on government-regulated utility emission standards.
- Sulfur dioxide, which is linked to acid rain, would increase slightly.
- Waste oil dumping would decrease, because EVs do not require crankcase oil.
- EVs reduce noise pollution, because they are quieter than ICEVs.
- Thermal pollution by large power plants would increase with increased EV usage.

EVs will considerably reduce the major causes of smog, substantially eliminate ozone depletion, and reduce greenhouse gases. With stricter SO_2 power plant emission standards, EVs would have little impact on SO_2 levels. Pollution reduction is the driving force behind EV usage.

1.3.3 CAPITAL AND OPERATING COST COMPARISON

The initial EV capital costs are higher than ICEV capital costs primarily due to the lack of mass production opportunities. However, EV capital costs are expected to decrease as volume increases. Capital costs of EVs easily exceed capital costs of ICEVs due to the cost of the battery. The power electronics stages are also expensive,

TABLE 1.2
EV and ICEV Efficiencies from Crude Oil to Traction Effort

ICEV	Efficiency (%)	
	Max.	Min.
Crude oil		
Refinery (petroleum)	90	85
Distribution to fuel tank	99	95
Engine	22	20
Transmission/axle	98	95
Wheels		
Overall efficiency (crude oil to wheels)	19	15

EV	Efficiency (%)	
	Max.	Min.
Crude oil		
Refinery (fuel oil)	97	95
Electricity generation	40	33
Transmission to wall outlet	92	90
Battery charger	90	85
Battery (lead/acid)	75	75
Motor/controller	85	80
Transmission/axle	98	95
Wheels		
Overall efficiency (crude oil to wheels)	20	14

although not at the same level as batteries. Total life cycle cost of an EV is projected to be less than that of a comparable ICEV. EVs are more reliable and will require less maintenance, giving a favorable bias over ICEV as far as operating cost is concerned.

1.3.4 U.S. DEPENDENCE ON FOREIGN OIL

The importance of searching for alternative energy sources cannot be overemphasized, and sooner or later, there will be another energy crisis if we, the people of the earth, do not reduce our dependence on oil. Today's industries, particularly the transportation industry, are heavily dependent on oil, the reserve of which will eventually deplete in the not so distant future. Today, about 42% of petroleum used for transportation in the United States is imported. An average ICEV in its lifetime uses 94 barrels of oil, based on 28 mi/gallon fuel consumption. On the other hand, an average EV uses two barrels of oil in its lifetime, based on

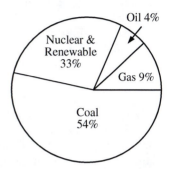

FIGURE 1.7 Electricity generation pie chart.

4 mi/kWh. The oil is used in the EV process during electricity generation, although only 4% of electricity generated is from oil. The energy sources for electricity generation are shown in the pie chart of Figure 1.7.

1.4 EV MARKET

We normally discuss the use of EVs for passenger and public transportation but tend to forget about their use as off-road vehicles in specialty applications, where range is not an issue. EVs have penetrated the market of off-road vehicles successfully over the years for clean air as well as for cost advantages. Examples of such applications are airport vehicles for passenger and ground support; recreational vehicles as in golf carts and for theme parks, plant operation vehicles like forklifts and loader trucks; vehicles for disabled persons; utility vehicles for ground transportation in closed but large compounds; etc. There are also EVs that run on tracks for material haulage in mines. There is potential for EV use for construction vehicles. The locomotives that run on tracks with electricity supplied from transmission lines are theoretically no different from other EVs, the major difference being in the way energy is fed for the propulsion motors.

Motivated by the growing concern about global pollution and the success of electric motor driven transportation in various areas, the interest is ever increasing for road EVs that can deliver the performance of ICEV counterparts. The major impediments for mass acceptance of EVs by the general public are the limited EV range and the lack of EV infrastructure. The solution of the range problem may come from extensive research and development efforts in batteries, fuel cells, and other alternative energy storage devices. An alternative approach is to create awareness among people on the problems of global warming and the advantages of EVs, while considering the fact that most people drive less than 50 miles a day, a requirement that can be easily met by today's technology.

The appropriate infrastructure must also be in place for EVs to become more popular. The issues related to infrastructure are as follows:

- Battery charging facilities: residential and public charging facilities and stations
- Standardization of EV plugs, cords, and outlets, and safety issues
- Sales and distribution
- Service and technical support
- Parts supply

The current initial cost of an EV is also a big disadvantage for the EV market. The replacement of the batteries, even for HEVs, is quite expensive, added to which is the limited life problem of these batteries. The cost of EVs will come down as volume goes up, but in the meantime, subsidies and incentives from the government can create momentum.

The increasing use of EVs will improve the job prospects of electrical engineers. The new jobs related to EVs will be in the following areas:

- *Power electronics and motor drives*: Design and development of the electrical systems of an EV
- *Power generation*: Increased utility demand due to EV usage
- *EV infrastructure*: Design and development of battery charging stations and of hydrogen generation, storage and distribution systems

REFERENCES

1. California Air Resources Board Office of Strategic Planning, Air-Pollution Transportation Linkage, 1989.
2. Wakefield, E.H., *History of Electric Automobile*, Society of Automotive Engineers, Warrendale, PA, 1994.
3. Westbrook, M.H., *The Electric Car*, The Institute of Electrical Engineers, London, United Kingdom, and Society of Automotive Engineers, Warrendale, PA, 2001.
4. Hodkinson, R. and Fenton, J., *Lightweight Electric/Hybrid Vehicle Design*, Society of Automotive Engineers, Warrendale, PA, 2001.
5. The Energy Foundation, 2001 annual report.

ASSIGNMENT

Search through reference materials and write a short report on the following topics:

1. Commercial and research EV/HEV programs around the world over the last 5 years, describing the various programs, goals, power range, motor used, type of IC engine, battery source, etc.
2. Case study of a recent EV/HEV
3. State and federal legislations and standardizations

2 Vehicle Mechanics

The fundamentals of vehicle design are embedded in the basic mechanics of physics, particularly in Newton's second law of motion relating force and acceleration. Newton's second law states that *the acceleration of an object is proportional to the net force exerted on it.* The object accelerates when the net force is nonzero, where the term "net force" refers to the result of the forces acting on the object. In the vehicle system, several forces act on it, with the resultant or net force dictating the motion according to Newton's second law. A vehicle moves forward with the aid of the force delivered by the propulsion unit overcoming the resisting forces due to gravity, air, and tire resistance. The acceleration and speed of the vehicle depend on the torque and power available from the traction unit and the existing road and aerodynamic conditions. Acceleration also depends on the composite mass of the vehicle, including the propulsion unit, all mechanical and electrical components, and the batteries.

A vehicle is designed based on certain given specifications and requirements. Furthermore, the electric and hybrid vehicle system is large and complex, involving multidisciplinary knowledge. The key to designing such a large system is to divide and conquer. The system-level perspective helps in mastering the design skills for a complex system, where the broad requirements are first determined and then system components are designed with more focused guidelines. For example, first, the power and energy requirements from the propulsion unit are determined from a given set of vehicle cruising and acceleration specifications. The component-level design begins in the second stage, where the propulsion unit, the energy source, and other auxiliary units are specified and designed. In this stage, the electrical and mechanical engineers start designing the electric motor for electric vehicles (EVs) or the combination of electric motor and internal combustion (IC) engine for hybrid electric vehicles (HEVs). The power electronics engineers design the power conversion unit that links the energy source with the electric motor. The controls engineer works in conjunction with the power electronics engineer to develop the propulsion control system. The chemists and the chemical engineers have the primary responsibility of designing the energy source based on the energy requirement and guidelines of the vehicle manufacturer. Many of the component designs proceed in an iterative manner, where various designers interact to ensure that the design goals are met.

Design is an iterative process, which starts with some known factors and other educated estimates or assumptions to be followed by scientific analysis in order to verify that the requirements are met. In this chapter, we will develop the tools for scientific analysis of vehicle mechanics based on Newton's second law of motion. After defining and describing a roadway, the vehicle kinetics issues will be addressed. The roadway and kinetics will be linked to establish the equation for the force required from the propulsion unit. The force from the propulsion unit, which can be an electric motor or an IC engine or a combination of the two, is known as

17

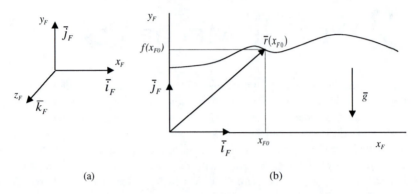

(a)   (b)

FIGURE 2.1 (a) Fixed coordinate system. (b) Roadway on the fixed coordinate system.

tractive force, F_{TR}. Once the force requirement is defined, one would proceed to calculate the power and energy required for a vehicle under consideration. The emphasis of study in this book is on the broad design goals, such as finding the power and energy requirements and predicting the range for a given energy source, thereby maintaining a top-level perspective. The design details of a subsystem are beyond the scope of this book, and readers are referred to the literature of the respective areas for further details.

2.1 ROADWAY FUNDAMENTALS

A vehicle moves on a level road and also up and down the slope of a roadway. We can simplify our description of the roadway by considering a straight roadway. Furthermore, we will define a tangential coordinate system that moves along with the vehicle, with respect to a fixed two-dimensional system. The roadway description will be utilized to calculate the distance traversed by a vehicle along the roadway.

The fixed coordinate system is attached to the earth such that the force of gravity is perpendicular to the unit vector $\bar{i}_F$ and in the $x_F y_F$ plane. Let us consider a straight roadway, i.e., the steering wheel is locked straight along the x_F direction. The roadway is then in the $x_F y_F$ plane of the fixed coordinate system (Figure 2.1).

The two-dimensional roadway can be described as $y_F = f(x_F)$. The *roadway position vector* $\bar{r}(x_F)$ between two points *a* and *b* along the horizontal direction is

$$\bar{r}(x_F) = x_F \bar{i}_F + f(x_F)\bar{j}_F \quad \text{for} \quad a \le x_F \le b$$

The direction of motion and the distance traversed by the vehicle are easier to express in terms of the *tangent vector* of the roadway position vector, given as:

$$\bar{T}(x_F) = \frac{d\bar{r}}{dx_F} = \bar{i}_F + \frac{df}{dx_F}\bar{j}_F$$

The distance-norm of the tangent vector $\left\| \bar{T}\left(x_F\right) \right\|$ is

$$\left\| \bar{T}\left(x_F\right) \right\| = \sqrt{1 + \left[\frac{df}{dx_F} \right]^2}$$

The tangential roadway length s is the distance traversed along the roadway. Mathematically, s is the arc length of $y_F = f(x_F)$ over $a \le x_F \le b$. Therefore,

$$s = \int_a^b \left\| \bar{T}\left(x_F\right) \right\| dx_F$$

The *roadway percent grade* is the vertical rise per 100 horizontal distance of roadway, with both distances expressed in the same units. The angle β of the roadway associated with the slope or grade is the angle between the tangent vector and the horizontal axis x_F (Figure 2.2). If Δy is the vertical rise in meters, then

$$\% \text{ grade} = \frac{\Delta y}{100\,\text{m}} 100\% = \Delta y \%$$

The tangent of the slope angle is

$$\tan \beta = \frac{\Delta y}{100\,\text{m}}$$

The percent grade, or β, is greater than zero when the vehicle is on an upward slope and is less than zero when the vehicle is going downhill. The roadway percent grade can be described as a function of the roadway, as

$$\beta\left(x_F\right) = \tan^{-1}\left[\frac{df\left(x_F\right)}{dx_F} \right]$$

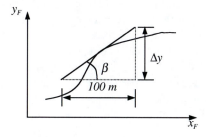

FIGURE 2.2 Grade of the roadway.

EXERCISE 2.1

A straight roadway has a profile in the $x_F y_F$-plane given by $f(x_F) = 3.9\sqrt{x_F}$ for $0 = x_F = 2$ miles (x_F and y_F are given in feet).

   (a) Plot the roadway.
   (b) Find $\beta(x_F)$.
   (c) Find the percent grade at $x_F = 1$ mile.
   (d) Find the tangential road length between 0 and 2 miles.

Solution

   (b) $\tan^{-1} \dfrac{1.95}{\sqrt{x_F}}$

   (c) 2.68%
   (d) 10,580 ft

2.2 LAWS OF MOTION

Newton's second law of motion can be expressed in equation form as follows:

$$\sum_i \overline{F}_i = m\overline{a}$$

where

$$\sum_i \overline{F}_i$$

is the net force, m is the effective mass, and $\overline{a}$ is the acceleration. The law is applied to the vehicle by considering a number of objects located at several points of contact of the vehicle with the outside world on which the individual forces act. Examples of such points of contact are the front and rear wheels touching the roadway surface and vehicle, the frontal area that meets the force from air resistance, etc. We will simplify the problem by merging all of these points of contact into one location at the center of gravity (cg) of the EV and HEV, which is justified, because the extent of the object is immaterial. For all of the force calculations to follow, we will consider the vehicle to be a particle mass located at the cg of the vehicle. The cg can be considered to be within the vehicle, as shown in Figure 2.3.

    Particle motion is described by particle velocity and acceleration characteristics. For the position vector $\overline{r}$ for the particle mass on which several forces are working, as shown in Figure 2.4, the velocity v and acceleration a are

$$\overline{v} = \frac{d\overline{r}}{dt} \quad \text{and} \quad \overline{a} = \frac{d\overline{v}}{dt}$$

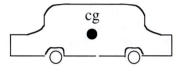

FIGURE 2.3 Center of gravity (cg) of a vehicle.

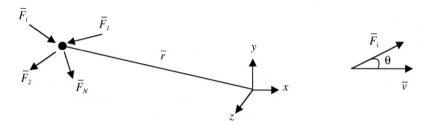

FIGURE 2.4 Forces on a particle.

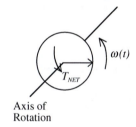

FIGURE 2.5 Rigid body rotation.

The power input to the particle for the ith force is

$$P_i = \bar{F}_i . \bar{v} = |\bar{F}_i| |\bar{v}| \cos\theta$$

where θ is the angle between F_i and the resultant velocity v.

For a rigid body rotating about a fixed axis, the equivalent terms relating motion and power are torque, angular velocity, and angular acceleration. Let there be i independent torques acting on a rigid body, causing it to rotate about an axis of rotation, as shown in Figure 2.5. If J (unit: kg-m^2) is the polar moment of inertia of the rigid body, then the rotational form of Newton's second law of motion is

$$\sum_i T_i = T_{NET} = J\alpha$$

where

$$\omega = \text{angular speed (rad/s)}$$

$$\alpha = \frac{d\omega}{dt} = \frac{d^2\theta}{dt^2} = \text{angular acceleration (rad/s}^2)$$

The power input for the ith torque is $P_i = T_i\omega$.

2.3 VEHICLE KINETICS

The tangential direction of forward motion of a vehicle changes with the slope of the roadway. To simplify the equations, a tangential coordinate system is defined below, so that the forces acting on the vehicle can be defined through a one-directional equation. Let $\bar{u}_T(x_F)$ be the *unit tangent vector* in the fixed coordinate system pointing in the direction of increasing x_F. Therefore,

$$\bar{u}_T(x_F) = \frac{\bar{T}_F(x_F)}{\|\bar{T}_F(x_F)\|} = \frac{\bar{i}_F + \dfrac{df}{dx_F}\bar{j}_F}{\sqrt{1 + \left[\dfrac{df}{dx_F}\right]^2}}$$

The tangential coordinate system shown in Figure 2.6 has the same origin as the fixed coordinate system. The z-direction unit vector is the same as that in the fixed coordinate system, but the x- and y-direction vectors constantly change with the slope of the roadway.

Newton's second law of motion can now be applied to the cg of EV in the tangential coordinate system as

$$\sum \bar{F}_T = m\bar{a}_T = m\frac{d\bar{v}_T}{dt}$$

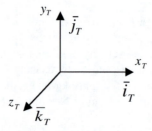

FIGURE 2.6 Tangential coordinate system and the unit tangent vector on a roadway.

where m is the effective vehicle mass. The components of the coordinate system are:

Component tangent to the road:

$$\sum \overline{F}_{xT} = m \frac{d\overline{v}_{xT}}{dt}$$

Component normal to the road:

$$\sum \overline{F}_{yT} = m \frac{d\overline{v}_{yT}}{dt}$$

And, because motion is assumed to be confined to the $x_T y_T$ plane:

$$\sum \overline{F}_{zT} = m \frac{d\overline{v}_{zT}}{dt} = 0$$

The vehicle tangential velocity is v_{xT}. The gravitational force in the normal direction is balanced by the road reaction force and, hence, there will be no motion in the y_T normal direction. In other words, the tire always remains in contact with the road. Therefore, the normal velocity $v_{yT} = 0$. Vehicle motion has been assumed confined to the $x_F y_F$ or $x_T y_T$ plane and, hence, there is no force or velocity acting in the z-direction. These justified simplifications allow us to use a one-directional analysis for vehicle propulsion in the x_T-direction. It is shown in the following that the vehicle tractive force and the opposing forces are all in the x_T-direction. Hence, the vector "-" will not be used in the symbols for simplicity.

The propulsion unit of the vehicle exerts a tractive force, F_{TR}, to propel the vehicle forward at a desired velocity. The tractive force must overcome the opposing forces, which are summed together and labeled as the *road load force*, F_{RL}. The road load force consists of the gravitational force, rolling resistance of the tires, and the aerodynamic drag force. The road load force is as follows:

$$F_{RL} = F_{gxT} + F_{roll} + F_{AD} \tag{2.1}$$

where x_T is the tangential direction along the roadway. The forces acting on the vehicle are shown in Figure 2.7.

The gravitational force depends on the slope of the roadway. The force is positive when climbing a grade and is negative when descending a downgrade roadway. The gravitational force to be overcome by the vehicle moving forward is

$$F_{gxT} = mg \sin \beta \tag{2.2}$$

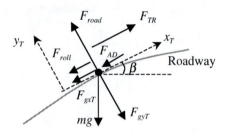

FIGURE 2.7 Forces acting on a vehicle.

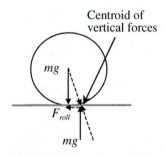

FIGURE 2.8 Rolling resistance force of wheels.

where m is the total mass of the vehicle, g is the gravitational acceleration constant, and β is the grade angle with respect to the horizon.

The rolling resistance is produced by the hysteresis of the tire at the contact surface with the roadway. In a stationary tire, the normal force to the road balances the distributed weight borne by the wheel at the point of contact along the vertical line beneath the axle. When the tire rolls, the centroid of the vertical forces on the wheel moves forward from beneath the axle toward the direction of motion of the vehicle, as shown in Figure 2.8. The weight on the wheel and the road normal force are misaligned due to the hysteresis of the tire. They form a couple that exerts a retarding torque on the wheel. The *rolling resistance force*, F_{roll}, is the force due to the couple, which opposes the motion of the wheel. The force F_{roll} is tangential to the roadway and always assists in braking or retarding the motion of the vehicle. The tractive force F_{TR} must overcome this force F_{roll} along with the gravitational force and the aerodynamic drag force. Rolling resistance can be minimized by keeping the tires as inflated as possible as well as by reducing the hysteresis. The ratio of the retarding force due to rolling resistance and the vertical load on the wheel is known as the *coefficient of rolling resistance*, C_0. The rolling resistance force is given by the following:

$$F_{roll} = \begin{cases} \text{sgn}[v_{xt}]mg\left(C_0 + C_1 v_{xT}^2\right) & \text{if} \quad v_{xT} \neq 0 \\ \left(F_{TR} - F_{gxT}\right) & \text{if} \quad v_{xT} = 0 \quad \text{and} \quad \left|F_{TR} - F_{gxT}\right| \leq C_0\, mg \\ \text{sgn}\left[F_{TR} - F_{gxT}\right]\left(C_0\, mg\right) & \text{if} \quad v_{xt} = 0 \quad \text{and} \quad \left|F_{TR} - F_{gxT}\right| > C_0\, mg \end{cases} \quad (2.3)$$

Typically, $0.004 < C_0 < 0.02$ (unitless), and $C_1 << C_0$ (s²/m²). $C_0 mg$ is the maximum rolling resistance at standstill. The sgn[v_{xT}] is the signum function given as

$$\text{sgn}\left[v_{xT}\right] = \begin{cases} 1 & \text{if} & v_{xT} \geq 0 \\ -1 & \text{if} & v_{xT} < 0 \end{cases}$$

The aerodynamic drag force is the result of viscous resistance and pressure distribution over the body of the air working against the motion of the vehicle. The force is given by

$$F_{AD} = \text{sgn}\left[v_{xT}\right]\left\{0.5\,\rho C_D A_F \left(v_{xT} + v_0\right)^2\right\} \tag{2.4}$$

where ρ is the air density in kg/m³, C_D is the aerodynamic drag coefficient (dimensionless, and typically is $0.2 < C_D < 0.4$), A_F is the equivalent frontal area of the vehicle, and v_0 is the head-wind velocity.

2.4 DYNAMICS OF VEHICLE MOTION

Tractive force is the force between the vehicle's tires and the road (and parallel to the road) supplied by the electric motor in an EV and by the combination of electric motor and IC engine in an HEV to overcome the road load. The dynamic equation of motion in the tangential direction is given by

$$k_m m \frac{dv_{xT}}{dt} = F_{TR} - F_{RL} \tag{2.5}$$

where k_m is the rotational inertia coefficient to compensate for the apparent increase in the vehicle's mass due to the onboard rotating mass. The typical values of k_m are between 1.08 and 1.1, and it is dimensionless. The acceleration of the vehicle is dv_{xT}/dt.

Dynamic equations can be represented in the state space format for simulation of an EV or HEV system. The motion described by Equation 2.5 is the fundamental equation required for dynamic simulation of the vehicle system. One of the state variables of the vehicle dynamical system is v_{xT}. The second equation needed for modeling and simulation is the velocity equation, where either s or x_F can be used as the state variable. The slope of the roadway β will be an input to the simulation model, which may be given in terms of the tangential roadway distance as $\beta = \beta(s)$ or in terms of the horizontal distance as $\beta = \beta(x_F)$. If β is given in terms of s, then the second state variable equation is

$$\frac{ds}{dt} = v_{xT} \tag{2.6}$$

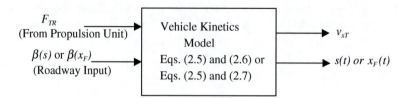

FIGURE 2.9 Modeling of vehicle kinetics and roadway.

If β is given in terms of x_F, then the second state variable equation is

$$\frac{dx_F}{dt} = \frac{v_{xT}}{\sqrt{1+\left[\dfrac{df}{dx_F}\right]^2}} \qquad (2.7)$$

The input–output relational diagram for simulating vehicle kinetics is shown in Figure 2.9.

2.5 PROPULSION POWER

The desired power rating of the electric motor or the power required from the combination of electric motor and IC engine (i.e., the propulsion unit) can be calculated from the above equations based on the system constraints of starting acceleration, vehicle rated and maximum velocity, and vehicle gradability. The torque at the wheels of the vehicle can be obtained from the power relation:

$$Power = T_{TR} \cdot \omega_{wh} = F_{TR} \cdot v_{xT} \text{ watts} \qquad (2.8)$$

where T_{TR} is the tractive torque in $N\text{-}m$, and ω_{wh} is the angular velocity of the wheel in rad/s. F_{TR} is in N, and v_{xT} is in m/s. Assuming no slip between the tires and the road, the angular velocity and the vehicle speed are related by

$$v_{xT} = \omega_{wh} \cdot r_{wh} \qquad (2.9)$$

where r_{wh} is the radius of the wheel in meters. The losses between the propulsion unit and wheels in the transmission and the differential have to be appropriately accounted for when specifying the power requirement of the propulsion unit.

An advantage of an electrically driven propulsion system is the elimination of multiple gears to match the vehicle speed and the engine speed. The wide speed range operation of electric motors enabled by power electronics control makes it possible to use a single gear-ratio transmission for instantaneous matching of the available motor torque T_{motor} with the desired tractive torque T_{TR}. The gear ratio and the size depend on the maximum motor speed, the maximum vehicle speed, the

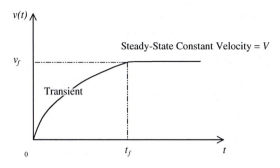

FIGURE 2.10 Plot of velocity profile.

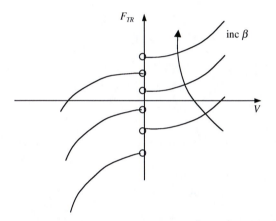

FIGURE 2.11 Tractive force versus steady state velocity characteristics.

wheel radius, and the available traction between the tires and the road. A higher
motor speed relative to the vehicle speed means a higher gear-ratio, larger size, and
higher cost. However, higher motor speed is also desired in order to increase the
power density of the motor. Therefore, a compromise is necessary between the
maximum motor speed and the gear-ratio to optimize the cost. Planetary gears are
typically used for EVs, with the gear-ratio rarely exceeding ten.

2.5.1 FORCE–VELOCITY CHARACTERISTICS

Having identified the fundamental forces and the associated dynamics for electric
and hybrid electric vehicles, let us now attempt to relate these equations to vehicle
design specifications and requirements. For an efficient design of the propulsion
unit, the designer must know the force required to accelerate the vehicle to a cruising
speed within a certain time and then to propel the vehicle forward at the rated steady
state cruising speed and at the maximum speed on a specified slope. Useful design
information is contained in the vehicle speed versus time and the steady state tractive
force versus constant velocity characteristics, illustrated in Figures 2.10 and 2.11.
In the sections to follow and in the remainder of the book, we will always assume

the velocity to be in the tangential direction and denote it by v instead of v_{xT} for simplicity. The steady state constant velocity will be denoted by the uppercase letter V.

Tractive force versus steady state velocity characteristics can be obtained from the equation of motion (Equation 2.5). When the steady state velocity is reached, $dv/dt = 0$; and, hence, $\Sigma F = 0$. Therefore, we have

$$F_{TR} - F_{AD} - F_{ROLL} - F_{gxT} = 0$$

$$\Rightarrow F_{TR} = mg\left[\sin\beta + C_0\,\text{sgn}(V)\right] + \text{sgn}(V)\left[mgC_1 + \frac{\rho}{2}C_D A_F\right]V^2$$

Note that

$$1.\ \frac{dF_{TR}}{dV} = 2V\,\text{sgn}(V)\left(\frac{\rho C_D A_F}{2} + mgC_1\right) > 0\ \forall V\ \text{ and}$$

$$2.\ \underset{V \to 0^+}{Lim}\ F_{TR} \neq \underset{V \to 0^-}{Lim}\ F_{TR}$$

The first equation reveals that the slope of F_{TR} versus V is always positive, meaning that the force requirement increases as the vehicle speed increases, which is primarily due to the aerodynamic drag force opposing vehicle motion. Also, discontinuity of the curves at zero velocity is due to the rolling resistance force.

2.5.2 MAXIMUM GRADABILITY

The maximum grade that a vehicle will be able to overcome with the maximum force available from the propulsion unit is an important design criterion as well as performance measure. The vehicle is expected to move forward very slowly when climbing a steep slope and, hence, we can make the following assumptions for maximum gradability:

1. The vehicle moves very slowly $\Rightarrow v \cong 0$.
2. F_{AD} and F_{roll} are negligible.
3. The vehicle is not accelerating, i.e., $dv/dt = 0$.
4. F_{TR} is the maximum tractive force delivered by a motor (or motors) at near zero speed.

At near stall conditions, under the above assumptions,

$$\Sigma F = 0 \Rightarrow F_{TR} - F_{gxT} = 0 \Rightarrow F_{TR} = mg\,\sin\beta$$

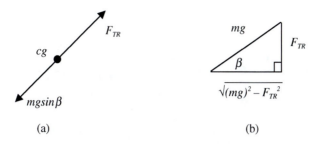

FIGURE 2.12 (a) Force diagram to determine maximum gradability. (b) Tractive force and *mg* with respect to the grade.

Therefore,

$$\sin\beta = \frac{F_{Tr}}{mg}.$$

The maximum percent grade is as follows (Figure 2.12):

$$\text{max \% grade} = 100\tan\beta$$

$$\Rightarrow \text{max \% grade} = \frac{100\,F_{TR}}{\sqrt{(mg)^2 - F_{TR}^2}} \qquad (2.10)$$

2.6 VELOCITY AND ACCELERATION

The energy required from the propulsion unit depends on the desired acceleration and the road load force that the vehicle has to overcome. Maximum acceleration is limited by the maximum tractive power available and the roadway condition at the time of vehicle operation. Although the road load force is unknown in a real-world roadway, significant insight about the vehicle velocity profile and energy requirement can be obtained through studies of assumed scenarios. Vehicles are typically designed with a certain objective, such as maximum acceleration on a given roadway slope in a typical weather condition. Discussed in the following are two simplified scenarios that will set the stage for designing EVs and HEVs.

2.6.1 CONSTANT F_{TR}, LEVEL ROAD

In the first case, we will assume a level road condition, where the propulsion unit for an EV exerts a constant tractive force. The level road condition implies that $\beta(s)$ = 0. We will assume that the EV is initially at rest, which implies $v(0) = 0$. The free body diagram at $t = 0$ is shown in Figure 2.13.

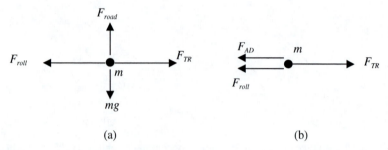

FIGURE 2.13 Forces acting on the vehicle on a level road. (a) Free body diagram $t = 0$. (b) Forces on the vehicle at $t > 0$.

Assume that $F_{TR}(0) = F_{TR} > C_0mg$, i.e., the initial tractive force is capable of overcoming the initial rolling resistance. Therefore,

$$\Sigma F(0) = ma(0) = m\frac{dv(0)}{dt}$$

$$\Rightarrow F_{TR} - C_0mg = m\frac{dv(0)}{dt}$$

Because $F_{TR}(0) > C_0mg$, $dv(0)/dt > 0$, and the velocity v increases. Once the vehicle starts to move, the forces acting on it change. At $t > 0$:

$$\Sigma F = m\frac{dv}{dt} \Rightarrow F_{TR} - F_{AD} - F_{roll} = m\frac{dv}{dt}$$

$$\Rightarrow F_{TR} - \text{sgn}[v(t)]\frac{\rho}{2}C_DA_Fv^2(t) - \text{sgn}[v(t)]mg(C_0 + C_1v^2(t)) = m\frac{dv}{dt}$$

Assuming $v(t) > 0$ for $t > 0$ and solving for dv/dt,

$$\frac{dv}{dt} = \left(\frac{F_{TR}}{m} - gC_0\right) - \left[\frac{\rho}{2m}C_DA_F + gC_1\right]v^2$$

Let us define the following constants for a constant F_{TR} acceleration:

$$K_1 = \frac{F_{TR}}{m} - gC_0 > 0$$

$$K_2 = \frac{\rho}{2m}C_DA_F + gC_1 > 0$$

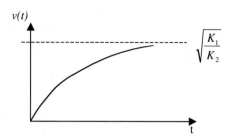

FIGURE 2.14 Velocity profile for a constant F_{TR} on a level road.

which gives

$$\frac{dv}{dt} = K_1 - K_2 v^2$$

2.6.1.1 Velocity Profile

The velocity profile for the constant F_{TR} level road case (Figure 2.14) can be obtained by solving for v from the dv/dt equation above, which gives

$$v(t) = \sqrt{\frac{K_1}{K_2}} \tanh\left(\sqrt{K_1 K_2}\, t\right) \tag{2.11}$$

The terminal velocity can be obtained by taking the limit of the velocity profile as time approaches infinity. The terminal velocity is

$$V_T = \lim_{t \to \infty} v(t) = \sqrt{\frac{K_1}{K_2}} \Rightarrow \sqrt{K_1 K_2} = K_2 V_T$$

2.6.1.2 Distance Traversed

The distance traversed by the vehicle can be obtained from the following relation:

$$\frac{ds(t)}{dt} = v(t) = V_T \tanh\left(K_2 V_T t\right)$$

The distance as a function of time is obtained by integrating the above equation:

$$s(t) = \frac{1}{K_2} \ln\left[\cosh\left(K_2 V_T t\right)\right] \tag{2.12}$$

The starting acceleration is often specified as 0 to v_f m/s in t_f s, where v_f is the desired velocity at the end of the specified time t_f s. The time to reach the desired velocity is given by

$$t_f = \frac{1}{K_2 V_T} \cosh^{-1}\left[e^{(K_2 s_f)}\right]$$
(2.13)

and the distance traversed during the time to reach the desired velocity is given by

$$s_f = \frac{1}{K_2} \ln\left[\cosh\left(K_2 V_T t_f\right)\right]$$
(2.14)

The desired time can also be expressed as follows:

$$t_f = \frac{1}{\sqrt{K_1 K_2}} \tanh^{-1}\left(\sqrt{\frac{K_2}{K_1}} v_f\right)$$
(2.15)

where v_f is the velocity after time t_f.

Let t_{V_T} = time to reach 98% of the terminal velocity V_T. Therefore,

$$t_{V_T} = \frac{1}{\sqrt{K_1 K_2}} \tanh^{-1}\left(\frac{0.98 V_T}{V_T}\right) \quad \text{or}$$

$$t_{V_T} = \frac{2.3}{\sqrt{K_1 K_2}} \quad \text{or} \quad \frac{2.3}{K_2 V_T}$$
(2.16)

2.6.1.3 Tractive Power

The instantaneous tractive power delivered by the propulsion unit is

$$P_{TR}(t) = F_{TR} v(t)$$

Substituting for $v(t)$,

$$P_{TR}(t) = F_{TR} V_T \tanh\left(K_2 V_T t\right)$$

$$\Rightarrow P_{TR}(t) = F_{TR} V_T \tanh\left(\sqrt{K_1 K_2} t\right) = P_T \tanh\left(\sqrt{K_1 K_2} t\right)$$
(2.17)

Terminal power can be expressed as $P_T = F_{TR} V_T$. The tractive power required to reach the desired velocity v_f over the acceleration interval Δt (refer to Figure 2.15) is

$$P_{TR,pk} = P_T \tanh\left(\sqrt{K_1 K_2} t_f\right)$$
(2.18)

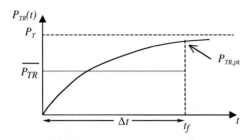

FIGURE 2.15 Acceleration interval $\Delta t = t_f - 0$.

The mean tractive power over the acceleration interval Δt is

$$\overline{P_{TR}} = \frac{1}{t_f} \int_0^{t_f} P_{TR}(t)\, dt$$

(2.19)

$$\Rightarrow \overline{P_{TR}} = \frac{P_T}{t_f} \frac{1}{\sqrt{K_1 K_2}} \ln\left[\cosh\left(\sqrt{K_1 K_2}\, t_f\right)\right]$$

2.6.1.4 Energy Required

The energy requirement for a given acceleration and constant steady state velocity is necessary for the design and selection of the energy source or batteries to cover a certain distance. The rate of change of energy is the tractive power, given as

$$P_{TR}(t) = \frac{de_{TR}}{dt}$$

where e_{TR} is the instantaneous tractive energy. The energy required or energy change during an interval of the vehicle can be obtained from the integration of the instantaneous power equation as

$$\int_{e_{TR}(0)}^{e_{TR}(t_f)} de_{TR} = \int_{t=0}^{t_f} P_{TR}\, dt$$

(2.20)

$$\Rightarrow \Delta e_{TR} = t_f \overline{P_{TR}}$$

EXAMPLE 2.1

An electric vehicle has the following parameter values:

$$m = 800 \text{ kg}, \ C_D = 0.2, \ A_F = 2.2 \text{ m}^2, \ C_0 = 0.008, \ C_1 = 1.6 * 10^{-6} \text{ s}^2/\text{m}^2$$

Also, take density of air $\rho = 1.18$ kg/m³, and acceleration due to gravity $g = 9.81$ m/s².

The vehicle is on level road. It accelerates from 0 to 65 mph in 10 s, such that its velocity pro le is gi ven by

$$v(t) = 0.29055t^2 \quad \text{for} \quad 0 \le t \le 10 \text{ s}$$

(a) Calculate $F_{TR}(t)$ for $0 \le t \le 10$ s.
(b) Calculate $P_{TR}(t)$ for $0 \le t \le 10$ s.
(c) Calculate the energy loss due to nonconservative forces E_{loss}.
(d) Calculate Δe_{TR}.

Solution

(a) From the force balance equation,

$$F_{TR} - F_{AD} - F_{roll} = m\frac{dv}{dt}$$

$$\Rightarrow F_{TR}(t) = m\frac{dv}{dt} + \frac{\rho}{2}C_D A_F v^2 + mg\left(C_0 + C_1 v^2\right)$$

$$= 464.88t + .02192\,t^4 + 62.78 \ N$$

(b) The instantaneous power is

$$P_{TR}(t) = F_{TR}(t) * v(t)$$

$$= 135.07t^3 + .00637t^6 + 18.24t^2 W$$

(c) The energy lost due to nonconservative forces is

$$E_{loss} = \int_0^{10} v\left(F_{AD} + F_{roll}\right)dt = \int_0^{10} 0.29055t^2\left(0.0219t^4 + 62.78\right)dt$$

$$= 15{,}180 \ J$$

(d) The kinetic energy of the vehicle is

$$\Delta KE = \frac{1}{2}m\left[v(10)^2 - v(0)^2\right] = 337{,}677 \ J$$

Therefore, the change in tractive energy is

$$\Delta e_{TR} = 15{,}180 + 337{,}677$$

$$= 352{,}857 \ J$$

EXERCISE 2.2

An EV has the following parameter values: $\rho = 1.16$ kg/m³, $m = 692$ kg, $C_D = 0.2$, $A_F = 2$ m², $g = 9.81$ m/s², $C_0 = 0.009$, and $C_1 = 1.75 * 10^{-6}$ s²/m². The EV undergoes constant F_{TR} acceleration on a level road starting from rest at $t = 0$. The maximum continuous F_{TR} that the electric motor is capable of delivering to the wheels is 1500 N.

(a) Find V_T (F_{TR}), and plot it.
(b) If $F_{TR} = 350$ N:
    (i) Find V_T.
    (ii) Plot $v(t)$ for $t \geq 0$.
    (iii) Find t_{V_T}.
    (iv) Calculate the time required to accelerate from 0 to 60 mph.
    (v) Calculate $P_{TR,pk}$, $\overline{P_{TR}}$, and Δe_{TR} corresponding to acceleration to $0.98 V_T$.

Solution

(a) $\quad V_T\left(F_{TR}\right) = 53.2\sqrt{1.45\times10^{-3}\ F_{TR} - 0.0883}$ m/s
(b)

    (i) 34.4 m/s
    (ii) $v(t) = 34.4 \tanh(1.22 \times 10^{-2}t)$ m/s
    (iii) 189 s
    (iv) 85.6 s
    (v) $P_{TR,pk} = 11.8$ kW, $\overline{P_{TR}} = 8.46$ kW, and $\Delta e_{TR} = 1.61$ MJ

2.6.2 NONCONSTANT F_{TR}, GENERAL ACCELERATION

In the general case, with a nonconstant F_{TR} and an arbitrary velocity pro le, as sho wn in Figure 2.16, force can be calculated as follows:

$$\Sigma F = m\frac{dv}{dt}$$

$$\Rightarrow F_{TR} - F_{AD} - F_{roll} - F_{gxT} = m\frac{dv}{dt}$$

$$\Rightarrow F_{TR} = m\frac{dv}{dt} + mg \sin\beta + F_{AD} + F_{roll} \qquad (2.21)$$

$$= m\frac{dv}{dt} + mg\sin\beta + \left[mgC_1 + \frac{\rho}{2}A_FC_D\right]v^2 + mgC_0$$

The instantaneous tractive power $P_{TR}(t)$ is

$$P_{TR}(t) = F_{TR}(t)v(t)$$

$$= mv\frac{dv}{dt} + v\left(F_{gxT} + F_{AD} + F_{roll}\right) \qquad (2.22)$$

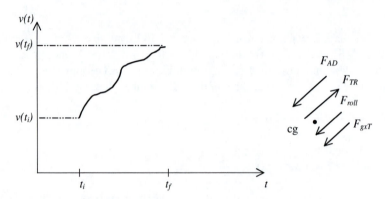

FIGURE 2.16 Arbitrary velocity profile.

The change in tractive energy Δe_{TR} is

$$\Delta e_{TR} = \int_{t_i}^{t_f} P_{TR}(t)\,dt$$

$$= m \int_{v(t_i)}^{v(t_f)} v\,dv + \int_{t_i}^{t_f} (v)\,F_{gxT}\,dt + \int_{t_i}^{t_f} (v)\big(F_{AD} + F_{roll}\big)\,dt$$

(2.23)

The energy supplied by the propulsion unit is converted into various forms of energy, some of which are stored in the vehicle system, while others are lost due to non-constructive forces. It is interesting to note the type of energy associated with each term in Equation 2.23. Let us consider the first term on the right side of Equation 2.23:

$$m \int_{v(t_i)}^{v(t_f)} v\,dv = \frac{1}{2} m\big[v^2(t_f) - v^2(t_i)\big] = \Delta(\text{Kinetic Energy})$$

Also,

$$\int_{t_i}^{t_f} (v)\,F_{gxT}\,dt = mg\int_{t_i}^{t_f} v\sin\beta\,dt = mg\int_{s(t_i)}^{s(t_f)} \sin\beta\,ds = mg\int_{f(t_i)}^{f(t_f)} df$$

$$= mg\big[f(t_f) - f(t_i)\big]$$

$$= \Delta(\text{Potential Energy})$$

The above term represents change in vertical displacement multiplied by mg, which is the change in potential energy.

The third and fourth terms on the right side of Equation 2.23 represent the energy required to overcome nonconstructive forces that include rolling resistance and aerodynamic drag force. The energy represented in these terms is essentially the loss term. Therefore,

$$\int_{t_i}^{t_f} (v)\left(F_{AD} + F_{roll}\right) dt = E_{loss}$$

Let

$$K_3 = mg\, C_0, \quad K_4 = mg\, C_1 + \frac{\rho}{2} C_D A_F$$

For

$$v(t) > 0,\ t_i \leq t \leq t_f,$$

$$E_{loss} = K_3 \int_{t_i}^{t_f} \frac{ds}{dt}\, dt + K_4 \int_{t_i}^{t_f} v^3\, dt;$$

$$= K_3 \Delta s + K_4 \int_{t_i}^{t_f} v^3\, dt$$

In summary, we can write

$$\Delta e_{TR} = \frac{1}{2} m \left[v^2\left(t_f\right) - v^2\left(t_i\right) \right] + mg\left[f\left(t_f\right) - f\left(t_i\right) \right] + \int_{t_i}^{t_f} (v)\left(F_{AD} + F_{roll}\right) dt$$

or

$$\Delta e_{TR} = \Delta\left(\text{Kinetic Energy}\right) + \Delta\left(\text{Potential Energy}\right) + E_{loss}$$

EXERCISE 2.3

The vehicle with parameters as given in Exercise 2.2 accelerates from 0 to 60 mi/h in 13.0 s for the two following acceleration types: constant F_{TR} and uniform acceleration.

(a) Plot, on the same graph, the velocity profile of each acceleration type.
(b) Calculate and compare the tractive energy required for the two types of acceleration. $F_{TR} = $ const. $= 1548$ N

Solution

(b) $\Delta e_{TR} = 0.2752$ MJ for constant F_{TR}, and $\Delta e_{TR} = 0.2744$ MJ for uniform acceleration.

2.7 PROPULSION SYSTEM DESIGN

The steady state maximum velocity, maximum gradability, and velocity equations can be used in the design stage to specify the power requirement of a particular vehicle. The common design requirements related to vehicle power expected to be specified by a customer are the initial acceleration, rated velocity on a given slope, maximum % grade, and maximum steady state velocity.

The complete design is a complex issue involving numerous variables, constraints, considerations, and judgment, which is beyond the scope of this book. The basic equations and principles that lay the foundation for design of EVs and HEVs are presented in this book. Design considerations will be highlighted as and when appropriate. Let us consider the tractive power requirement for initial acceleration, which plays a significant role in determining the rated power of the propulsion unit. The initial acceleration is specified as 0 to v_f in t_f s. The design problem is to solve for F_{TR}, starting with a set of variables including vehicle mass, rolling resistance, aerodynamic drag coefficient, percent grade, wheel radius, etc., some of which are known, while others have to be assumed. The acceleration of the vehicle in terms of these variables is given by

$$a = \frac{dv}{dt} = \frac{F_{TR} - F_{RL}}{m} \tag{2.24}$$

The tractive force output of the electric motor for an EV or the combination of electric motor and IC engine for an HEV will be a function of vehicle velocity. Furthermore, the road load characteristics are a function of velocity, resulting in a transcendental equation to be solved to determine the desired tractive power from the propulsion unit. Other design requirements also play a significant role in determining tractive power. The problem is best handled by a computer simulation, with which the various equations of Sections 2.5, 2.6, and 2.7 can be used iteratively to calculate the tractive force and power requirements from the propulsion unit for the given set of specifications. Insight gained from the scenarios discussed in this chapter will be used later to specify the power ratings of the electric motor for EVs and HEVs in Chapter 9 and for IC engines for HEVs in Chapter 10.

PROBLEMS

2.1

A length of road is straight, and it has a pro le in the x–y plane described by

$$f(x) = 200 \ln\left[7.06 * 10^{-4}\left(x + 1416\right)\right]$$

where $0 \le x \le 3$ miles $= 15,840$ ft; $f(x)$ and x are in feet.

(a) Plot the road pro le in the x–y plane for $0 \le x \le 15,840$ ft.
(b) Derive an expression for $\beta(x)$. Calculate $\beta(500$ ft).
(c) Derive an expression for percent grade (x). Calculate percent grade (500 ft).
(d) Derive an expression for tangential road length $s(x)$, such that $s(0) = 0$. Calculate $s(500$ ft).
(e) Can you nd an e xpression for $x(s)$, i.e., can you express x as a function of s? Show some steps in your attempt.

2.2

A straight roadway has a pro le in the x–y plane gi ven by

$$f(x) = 4.1\sqrt{x_f} \quad \text{for} \quad 0 \le x_f \le 2 \text{ mile} = 10,560 \text{ ft}$$

$f(x_f)$ and x_f are measured in feet.

(a) Derive an expression for $\beta(x_f)$. Calculate $\beta(1$ mi).
(b) Calculate the tangential road length, s from 0 to 2 mi.

2.3

An electric vehicle has the following parameter values:

$$m = 692 \text{ kg}, \ C_D = 0.2, \ A_F = 2 \text{ m}^2, \ C_0 = 0.009, \text{ and } C_1 = 1.75 * 10^{-6} \text{ s}^2/\text{m}^2$$

Also, take $\rho = 1.16 \text{ kg/m}^3$ and $g = 9.81 \text{ m/s}^2$.

(a) *EV at rest* — The EV is stopped at a stop sign at a point in the road where the grade is +15%. The tractive force of the vehicle is supplied by the vehicle brakes.
   (i) Calculate the tractive force necessary for zero rolling resistance. (The vehicle is at rest.)
   (ii) Calculate the minimum tractive force required from the brakes to keep the EV from rolling down the grade.

(b) *EV at constant velocity* — The EV is moving at a constant velocity along a road that has a constant grade of –12%.

(i) Plot, on the same graph, the magnitudes of the tangential gravitational force (F_{gxT}), the aerodynamic drag force (F_{AD}), and the rolling resistance force (F_{roll}) versus velocity for $0 < V \le 180$ mph. Over that range of velocity, does F_{gxT} dominate? When does F_{AD} dominate? When does F_{roll} dominate? Label these regions on the graph.

(ii) Derive an expression for the tractive force as a function of velocity. Plot this expression on its own graph. Is the tractive force always in the same direction?

2.4

Showing all steps, derive and plot the velocity profile [i.e., $v(t)$] for constant F_{TR}-constant grade acceleration. (Constant grade means that β is constant but not necessarily zero.) Given:

1. The EV starts from rest at $t = 0$.
2. The resultant of F_{gxT} and F_{TR} is enough to overcome the rolling resistance to get the EV moving.
3. $v(t) \ge 0$ for $t \ge 0$.

What effect does gravity have on the velocity profile?

2.5

A vehicle accelerates from 0 to 60 mph in 10 s with the velocity profile given by

$$v(t) = 20 \ln(0.282t + 1) \text{ m/s} \quad \text{for} \quad 0 \le t \le 10 \text{ s}$$

The vehicle is on level road. For the problem, use the parameters given in Problem 2.3.

(a) Calculate and plot $F_{TR}(t)$ and $P_{TR}(t)$ for $0 \le t \le 10$ s
(b) Calculate Δe_{TR}. How much of Δe_{TR} is ΔKE? How much is E_{loss}?

2.6

Using the vehicle parameters given in Problem 2.3, calculate and plot, on the same graph, steady state F_{TR} versus V characteristics for $\beta = \pm 4°$ and -60 mph $\le V \le 60$ mph, but $V \ne 0$.

2.7

An electric vehicle racer will attempt to jump five city buses as shown in Figure P2.7. The vehicle will start at rest at a position 100 m from the beginning of the take-off ramp. The vehicle will accelerate uniformly, until it reaches the end of the take-off ramp, at which time it will be traveling at 100 mph.

The vehicle has the following parameter values:

$$m = 692 \text{ kg}, C_D = 0.2, A_F = 2 \text{ m}^2, C_0 = 0.009, \text{ and } C_1 = 1.75 * 10^{-6} \text{ s}^2/\text{m}^2$$

Also, take density of air $\rho = 1.16 \text{ kg/m}^3$, and acceleration due to gravity $g = 9.81 \text{ m/s}^2$.

(a) Sketch and label the velocity profile of the vehicle from the time it starts to the time it reaches the end of the take-off ramp. How much time does the vehicle take to reach the end of the take-off ramp?

(b) Calculate the change in gravitational potential energy, from the start to the end of the take-off ramp.

(c) Calculate energy loss, E_{loss}, from the start to the end of the take-off ramp, if $\Delta e_{TR} = 8.28 \times 10^5 \ J$ during that period.

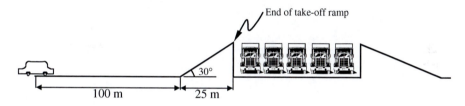

FIGURE P2.7

3 Energy Source: Battery

A basic requirement for electric vehicles (EVs) is a portable source of electrical energy, which is converted to mechanical energy in the electric motor for vehicle propulsion. Electrical energy is typically obtained through conversion of chemical energy stored in devices such as batteries and fuel cells. A flywheel is an alternative portable source in which energy is stored in mechanical form to be converted into electrical energy on demand for vehicle propulsion. The portable electrical energy source presents the biggest obstacle in commercialization of EVs. A near-term solution for minimizing the environmental pollution problem due to the absence of a suitable, high-energy-density energy source for EVs is perceived in the hybrid electric vehicles (HEVs) that combine propulsion efforts from gasoline engines and electric motors.

A comparison of the specific energy of the available energy sources is given in Table 3.1. The specific energy is the energy per unit mass of the energy source. The specific energies are shown without taking containment into consideration. The specific energy of hydrogen and natural gas would be significantly lower than that of gasoline when containment is considered.

Among the available choices of portable energy sources, batteries have been the most popular choice of energy source for EVs since the beginning of research and development programs in these vehicles. The EVs and HEVs commercially available today use batteries as the electrical energy source. The various batteries are usually compared in terms of descriptors, such as specific energy, specific power, operating life, etc. Similar to specific energy, *specific power* is the power available per unit mass from the source. The *operating life* of a battery is the number of deep discharge cycles obtainable in its lifetime or the number of service years expected in a certain application. The desirable features of batteries for EV and HEV applications are high specific power, high specific energy, high charge acceptance rate for recharging and regenerative braking, and long calendar and cycle life. Additional technical issues include methods and designs to balance the battery segments or packs electrically and thermally, accurate techniques to determine a battery's state of charge, and recycling facilities of battery components. And above all, the cost of batteries must be reasonable for EVs and HEVs to be commercially viable.

Battery technology has been undergoing extensive research and development efforts over the past 30 years, yet there is currently no battery that can deliver an acceptable combination of power, energy, and life cycle for high-volume production vehicles. The small number of EVs and HEVs that were introduced in the market used batteries that were too expensive and have short calendar life, making the batteries the biggest impediment in commercializing EVs and HEVs.

TABLE 3.1
Nominal Energy Density of Sources

Energy Source	Nominal Specific Energy (Wh/kg)
Gasoline	12,500
Natural gas	9350
Methanol	6050
Hydrogen	33,000
Coal (bituminous)	8200
Lead-acid battery	35
Lithium-polymer battery	200
Flywheel (carbon-fiber)	200

3.1 BATTERY BASICS

The batteries are made of unit cells containing the chemical energy that is convertible to electrical energy. One or more of these electrolytic cells are connected in series to form one battery. The grouped cells are enclosed in a casing to form a battery module. A battery pack is a collection of these individual battery modules connected in a series and parallel combination to deliver the desired voltage and energy to the power electronic drive system.

The energy stored in a battery is the difference in free energy between chemical components in the charged and discharged states. This available chemical energy in a cell is converted into electrical energy only on demand, using the basic components of a unit cell, which are the positive and negative electrodes, the separators, and the electrolytes. The electrochemically active ingredient of the positive or negative electrode is called the active material. Chemical oxidation and reduction processes take place at the two electrodes, thereby bonding and releasing electrons, respectively. The electrodes must be electronically conducting and are located at different sites, separated by a separator, as shown in Figure 3.1. During battery operation, chemical reactions at each of the electrodes cause electrons to flow from one electrode to another; however, the flow of electrons in the cell is sustainable only if electrons generated in the chemical reaction are able to flow through an external electrical circuit that connects the two electrodes. The connection points between the electrodes and the external circuit are called the battery terminals. The external circuit ensures that most of the stored chemical energy is released only on demand and is utilized as electrical energy. It must be mentioned that only in an ideal battery does current flow only when the circuit between the electrodes is completed externally. Unfortunately, many batteries do allow a slow discharge, due to diffusion effects, which is why they are not particularly good for long-term energy storage. This slow discharge with open-circuit terminals is known as *self-discharge*, which is also used as a descriptor of battery quality.

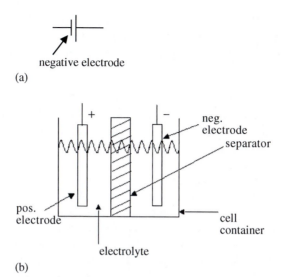

(a)

(b)

FIGURE 3.1 Components of a battery cell. (a) Cell circuit symbol; (b) cell cross-section.

The components of the battery cell are described as follows:

1. *Positive electrode*: The positive electrode is an oxide or sulfide or some other compound that is capable of being reduced during cell discharge. This electrode consumes electrons from the external circuit during cell discharge. Examples of positive electrodes are lead oxide (PbO_2) and nickel oxyhydroxide (NiOOH). The electrode materials are in the solid state.

2. *Negative electrode*: The negative electrode is a metal or an alloy that is capable of being oxidized during cell discharge. This electrode generates electrons in the external circuit during cell discharge. Examples of negative electodes are lead (Pb) and cadmium (Cd). Negative electrode materials are also in the solid state within the battery cell.

3. *Electrolyte*: The electrolyte is the medium that permits ionic conduction between positive and negative electrodes of a cell. The electrolyte must have high and selective conductivity for the ions that take part in electrode reactions, but it must be a nonconductor for electrons in order to avoid self-discharge of batteries. The electrolyte may be liquid, gel, or solid material. Also, the electrolyte can be acidic or alkaline, depending on the type of battery. Traditional batteries such as lead-acid and nickel-cadmium use liquid electrolytes. In lead-acid batteries, the electrolyte is the aqueous solution of sulfuric acid [H_2SO_4(aq)]. Advanced batteries currently under development for EVs, such as sealed lead-acid, nickel-metal-hydride (NiMH), and lithium-ion batteries use an electrolyte that is gel, paste, or resin. Lithium-polymer batteries use a solid electrolyte.

4. *Separator*: The separator is the electrically insulating layer of material that physically separates electrodes of opposite polarity. Separators must be permeable to the ions of the electrolyte and may also have the function of storing or immobilizing the electrolyte. Present day separators are made from synthetic polymers.

There are two basic types of batteries: primary batteries and secondary batteries. Batteries that cannot be recharged and are designed for a single discharge are known as primary batteries. Examples of these are the lithium batteries used in watches, calculators, cameras, etc., and the manganese dioxide batteries used to power toys, radios, torches, etc. Batteries that can be recharged by flowing current in the direction opposite to that during discharge are known as secondary batteries. The chemical reaction process during cell charge operation when electrical energy is converted into chemical energy is the reverse of that during discharge. The batteries needed and used for EVs and HEVs are all secondary batteries, because they are recharged during regeneration cycles of vehicle operation or during the battery recharging cycle in the stopped condition using a charger. All the batteries that will be discussed in the following are examples of secondary batteries.

The major types of rechargeable batteries considered for EV and HEV applications are:

- Lead-acid (Pb-acid)
- Nickel-cadmium (NiCd)
- Nickel-metal-hydride (NiMH)
- Lithium-ion (Li-ion)
- Lithium-polymer (Li-poly)
- Sodium-sulfur (NaS)
- Zinc-air (Zn-Air)

The lead-acid type of battery has the longest development history of all battery technology, particularly for their need and heavy use in industrial EVs, such as for golf carts in sports, passenger cars in airports, and forklifts in storage facilities and supermarkets. Research and development for batteries picked up momentum following the resurgence of interest in EVs and HEVs in the late 1960s and early 1970s. Sodium-sulfur batteries showed great promise in the 1980s, with high energy and power densities, but safety and manufacturing difficulties led to the abandonment of the technology. The development of battery technology for low-power applications, such as cell phones and calculators, opened the possibilities of scaling the energy and power of nickel-cadmium- and lithium-ion-type batteries for EV and HEV applications.

The development of batteries is directed toward overcoming significant practical and manufacturing difficulties. Theoretical predictions are difficult to match in manufactured products due to practical limitations. Theoretical and practical specific energies of several batteries are given in Table 3.2 for comparison.

The characteristics of some of the more important battery technologies mentioned above are given in the following. The theoretical aspects of the lead-acid

TABLE 3.2
Specific Energy of Batteries

Battery	Specific Energy (Wh/kg)	
	Theoretical	Practical
Lead-acid	108	50
Nickel-cadmium		20–30
Nickel-zinc		90
Nickel-iron		60
Zinc-chlorine		90
Silver-zinc	500	100
Sodium-sulfur	770	150–300
Aluminum-air		300

battery will be discussed in detail rst, follo wed by shorter descriptions of the other promising technologies.

3.2 LEAD-ACID BATTERY

Lead-acid batteries have been the most popular choice of batteries for EVs. Lead-acid batteries can be designed to be high powered and are inexpensive, safe, and reliable. A recycling infrastructure is in place for them. However, low speci c ener gy, poor cold temperature performance, and short calendar and cycle life are among the obstacles to their use in EVs and HEVs.

The lead-acid battery has a history that dates to the middle of the 19th century, and it is currently a mature technology. The rst lead-acid battery w as produced as early as in 1859. In the early 1980s, over 100,000,000 lead-acid batteries were produced per year. The long existence of the lead-acid battery is due to the following:

- Relatively low cost
- Easy availability of raw materials (lead, sulfur)
- Ease of manufacture
- Favorable electrochemical characteristics

The battery cell operation consists of a cell discharge operation, when the energy is supplied from the battery to the electric motor to develop propulsion power, and a cell charge operation, when energy is supplied from an external source to store energy in the battery.

3.2.1 CELL DISCHARGE OPERATION

In the cell discharge operation (Figure 3.2), electrons are consumed at the positive electrode, the supply of which comes from the negative electrode. The current o w is, therefore, out of the positive electrode into the motor-load, with the battery acting as the source.

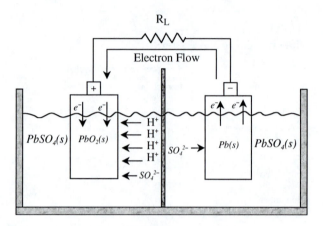

FIGURE 3.2 Lead-acid battery: cell discharge operation.

The positive electrode equation is given by:

$$PbO_2(s) + 4H^+(aq) + SO_4^{2-}(aq) + 2e \rightarrow PbSO_4 + 2H_2O(l)$$

A highly porous structure is used for the positive electrode to increase the PbO_2(s)/electrolyte contact area, which is about 50 to 150 m² per Ah of battery capacity. This results in higher current densities, as PbO_2 is converted to $PbSO_4$(s). As discharge proceeds, the internal resistance of the cell rises due to $PbSO_4$ formation and decreases the electrolyte conductivity as H_2SO_4 is consumed. $PbSO_4$(s) deposited on either electrode in a dense, fine-grain form can lead to sulfatation. The discharge reaction is largely inhibited by the buildup of $PbSO_4$, which reduces cell capacity significantly from the theoretical capacity.

The negative electrode equation during cell discharge is:

$$Pb(s) + SO_4^{2-}(aq) \rightarrow PbSO_4 + 2e$$

The electrons are released at the negative electrode during discharge operation. The production of PbSO4(s) can degrade battery performance by making the negative electrode more passive.

The overall cell discharge chemical reaction is:

$$Pb(s) + PbO_2(s) + 2H_2SO_4(aq) \rightarrow 2PbSO_4 + 2H_2O(l)$$

3.2.2 CELL CHARGE OPERATION

The cell charge operation (Figure 3.3) is the reverse of the cell discharge operation. During cell charging, lead sulfate is converted back to the reactant states of lead and lead oxide. The electrons are consumed from the external source at the negative electrode, while the positive electrode releases the electrons. The current flows into the positive electrode from the external source, thereby delivering electrical energy

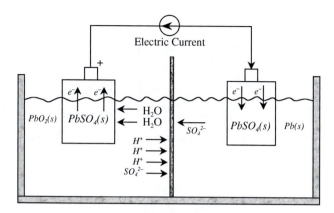

FIGURE 3.3 Lead-acid battery: cell charge operation.

into the cell, where it gets converted into chemical energy. The chemical reaction at the positive electrode during cell charging is:

$$PbSO_4(s) + 2H_2O(l) \rightarrow PbO_2(s) + 4H^+(aq) + SO_4^{2-}(aq) + 2e$$

The chemical reaction at the negative electrode during cell charging is:

$$PbSO_4(s) + 2e \rightarrow Pb(s) + SO_4^{2-}(aq)$$

The overall chemical reaction during cell charging is:

$$2PbSO_4(s) + 2H_2O(l) \rightarrow Pb(s) + PbO_2(s) + 2H_2SO_4(aq)$$

Conventionally, lead-acid batteries are of flooded-electrolyte cells, where free acid covers all the plates. This imposes the constraint of maintaining an upright position for the battery, which is difficult in certain portable situations. Efforts in developing hermetically sealed batteries faced the problem of buildup of an explosive mixture of hydrogen and oxygen on approaching the top-of-charge or overcharge condition during cell recharging. The problem is addressed in the valve-regulated-lead-acid (VRLA) batteries by providing a path for the oxygen, liberated at the positive electrode, to reach the negative electrode, where it recombines to form lead-sulfate. There are two mechanisms for making sealed VRLA batteries, the gel battery, and the AGM (absorptive glass microfiber) battery. These types are based on immobilizing the sulfuric acid electrolyte in the separator and the active materials, leaving sufficient porosity for the oxygen to diffuse through the separator to the negative plate.[1]

3.2.3 CONSTRUCTION

Construction of a typical battery consists of positive and negative electrode groups (elements) interleaved to form a cell. The through partition connection in the battery

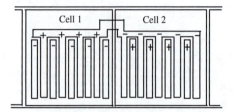

FIGURE 3.4 Schematic diagram of a lead-acid battery showing through-partition connection.

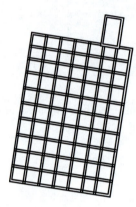

FIGURE 3.5 A lead-acid battery grid.

is illustrated in Figure 3.4. The positive plate is made of stiff paste of the active material on a lattice-type grid, which is shown in Figure 3.5. The grid, made of a suitably selected lead alloy, is the framework of a portable battery to hold the active material. The positive plates can be configured in flat pasted or tubular fashion. The negative plates are always manufactured as pasted types.

3.3 ALTERNATIVE BATTERIES

3.3.1 NICKEL-CADMIUM BATTERY

Nickel-cadmium (NiCd) and nickel-metal-hydride (NiMH) batteries are examples of alkaline batteries with which electrical energy is derived from the chemical reaction of a metal with oxygen in an alkaline electrolyte medium. The specific energy of alkaline batteries is lowered due to the addition of weight of the carrier metal. The NiCd battery employs a nickel oxide positive electrode and a metallic cadmium negative electrode. The net reaction occurring in the potassium hydroxide (KOH) electrolyte is:

$$Cd + 2NiOOH + 2H_2O \underset{\text{\textit{Charge}}}{\overset{\text{\textit{Discharge}}}{\underset{\leftarrow}{\rightarrow}}} 2Ni(OH)_2 + Cd(OH)_2 \text{ (1.299 volts)}$$

The practical cell voltage is 1.2 to 1.3 V, and the atomic mass of cadmium is 112. The specific energy of NiCd batteries is 30 to 50 Wh/kg, which is similar to that of lead-acid batteries. The advantages of NiCd batteries are superior low-temperature performance compared to lead-acid batteries, flat discharge voltage, long life, and excellent reliability. The maintenance requirements of the batteries are also low. The biggest drawbacks of NiCd batteries are the high cost and the toxicity contained in cadmium. Environmental concerns may be overcome in the long run through efficient recycling, but the insufficient power delivered by the NiCd batteries is another important reason for not considering these batteries for EV and HEV applications. The drawbacks of the NiCd batteries led to the rapid development of NiMH batteries, which are deemed more suitable for EV and HEV applications.

3.3.2 NICKEL-METAL-HYDRIDE (NiMH) BATTERY

The nickel-metal-hydride battery is a successor to the nickel-hydrogen battery and is already in use in production HEVs. In NiMH batteries, the positive electrode is a nickel oxide similar to that used in a NiCd battery, while the negative electrode is a metal hydride where hydrogen is stored. The concept of NiMH batteries is based on the fact that fine particles of certain metallic alloys, when exposed to hydrogen at certain pressures and temperatures, absorb large quantities of the gas to form the metal-hydride compounds. Furthermore, the metal hydrides are able to absorb and release hydrogen many times without deterioration. The two electrode chemical reactions in a NiMH battery are:

At the positive electrode,

$$NiOOH + H_2O + e^- \quad \overset{Discharge}{\underset{Charge}{\rightleftarrows}} \quad Ni(OH)_2 + OH^-$$

At the negative electrode,

$$MH_x + OH^- \quad \overset{Discharge}{\underset{Charge}{\rightleftarrows}} \quad MH_{x-1} + H_2O + e^-$$

M stands for metallic alloy, which takes up hydrogen at ambient temperature to form the metal hydride MH_x. The negative electrode consists of a compressed mass of fine metal particles. The proprietary alloy formulations used in NiMH are known as AB_5 and AB_2 alloys. In the AB_5 alloy, A is the mixture of rare earth elements, and B is partially substituted nickel. In the AB_2 alloy, A is titanium or zirconium, and B is again partially substituted nickel. The AB_2 alloy has a higher capacity for hydrogen storage and is less costly. The operating voltage of NiMH is almost the same as that of NiCd, with flat discharge characteristics. The capacity of the NiMH is significantly higher than that of NiCd, with specific energy ranging from 60 to 80 Wh/kg. The specific power of NiMH batteries can be as high as 250 W/kg.

The NiMH batteries have penetrated the market in recent years at an exceptional rate. The Chrysler electric minivan "Epic" uses a NiMH battery pack, which gives a range of 150 km. In Japan, NiMH battery packs produced by Panasonic EV Energy are being used in Toyota EV RAV-EV and Toyota HEV Prius. The components of NiMH are recyclable, but a recycling structure is not yet in place. NiMH batteries have a much longer life cycle than lead-acid batteries and are safe and abuse tolerant. The disadvantages of NiMH batteries are the relatively high cost, higher self-discharge rate compared to NiCd, poor charge acceptance capability at elevated temperatures, and low cell efficiency. NiMH is likely to survive as the leading rechargeable battery in the future for traction applications, with strong challenge coming only from lithium-ion batteries.[2]

3.3.3 LI-ION BATTERY

Lithium metal has high electrochemical reduction potential (3.045 V) and the lowest atomic mass (6.94), which shows promise for a battery of 3 V cell potential when combined with a suitable positive electrode. The interest in secondary lithium cells soared soon after the advent of lithium primary cells in the 1970s, but the major difficulty was the highly reactive nature of the lithium metal with moisture, restricting the use of liquid electrolytes. Discovery in the late 1970s by researchers at Oxford University that lithium can be intercalated (absorbed) into the crystal lattice of cobalt or nickel to form $LiCoO_2$ or $LiNiO_2$ paved the way toward the development of Li-ion batteries.[3] The use of metallic-lithium is bypassed in Li-ion batteries by using lithium intercalated (absorbed) carbons (Li_xC) in the form of graphite or coke as the negative electrode, along with the lithium metallic oxides as the positive electrode. The graphite is capable of hosting lithium up to a composition of LiC_6. The majority of the Li-ion batteries uses positive electrodes of cobalt oxide, which is expensive but proven to be the most satisfactory. The alternative positive electrode is based on nickel oxide $LiNiO_2$, which is structurally more complex but costs less. Performance is similar to that of cobalt oxide electrodes. Manganese-oxide-based positive electrodes ($LiMn_2O_4$ or $LiMnO_2$) are also under research, because manganese is cheaper, widely available, and less toxic.

The cell discharge operation in a lithium ion cell using $LiCoO_2$ is illustrated in Figure 3.6. During cell discharge, lithium ions (Li^+) are released from the negative electrode that travels through an organic electrolyte toward the positive electrode. In the positive electrode, the lithium ions are quickly incorporated into the lithium compound material. The process is completely reversible. The chemical reactions at the electrodes are as follows:

At the negative electrode,

$$Li_xC_6 \quad \underset{Charge}{\overset{Discharge}{\rightleftarrows}} \quad 6C + xLi^+ + xe^- \quad \text{where } 0 < x < 1$$

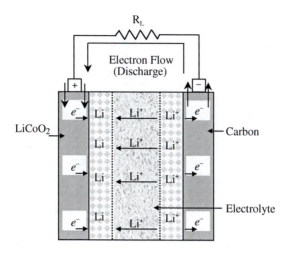

FIGURE 3.6 Lithium-ion cell.

At the positive electrode,

$$xLi^+ + xe^- + Li_{(1-x)}CoO_2 \quad \overset{Discharge}{\underset{Charge}{\overset{\rightarrow}{\leftarrow}}} \quad LiCoO_2$$

During cell charge operation, lithium ions move in the opposite direction from the positive electrode to the negative electrode. The nominal cell voltage for a Li-ion battery is 3.6 V, which is equivalent to three NiMH or NiCd battery cells.

Lithium-ion batteries have high specific energy, high specific power, high energy efficiency, good high-temperature performance, and low self-discharge. The components of Li-ion batteries are also recyclable. These characteristics make Li-ion batteries highly suitable for EV and HEV and other applications of rechargeable batteries.

3.3.4 LI-POLYMER BATTERY

Lithium-polymer evolved out of the development of solid state electrolytes, i.e., solids capable of conducting ions but that are electron insulators. The solid state electrolytes resulted from research in the 1970s on ionic conduction in polymers. These batteries are considered solid state batteries, because their electrolytes are solids. The most common polymer electrolyte is polyethylene oxide compounded with an appropriate electrolyte salt.

The most promising positive electrode material for Li-poly batteries is vanadium oxide V_6O_{13}.[1] This oxide interlaces up to eight lithium atoms per oxide molecule with the following positive electrode reaction:

$$\text{Li}_x + \text{V}_6\text{O}_{13} + x\text{e}^- \quad \overset{\textit{Discharge}}{\underset{\textit{Charge}}{\overset{\rightarrow}{\leftarrow}}} \quad \text{Li}_x\text{V}_6\text{O}_{13} \quad \text{where } 0 < x < 8$$

Li-poly batteries have the potential for the highest specific energy and power. The solid polymers, replacing the more flammable liquid electrolytes in other type of batteries, can conduct ions at temperatures above 60°C. The use of solid polymers also has a great safety advantage in case of EV and HEV accidents. Because the lithium is intercalated into carbon electrodes, the lithium is in ionic form and is less reactive than pure lithium metal. The thin Li-poly cell gives the added advantage of forming a battery of any size or shape to suit the available space within the EV or HEV chassis. The main disadvantage of the Li-poly battery is the need to operate the battery cell in the temperature range of 80 to 120°C. Li-poly batteries with high specific energy, initially developed for EV applications, also have the potential to provide high specific power for HEV applications. The other key characteristics of the Li-poly are good cycle and calendar life.

3.3.5 ZINC-AIR BATTERY

Zinc-air batteries have a gaseous positive electrode of oxygen and a sacrificial negative electrode of metallic zinc. The practical zinc-air battery is only mechanically rechargeable by replacing the discharged product, zinc hydroxide, with fresh zinc electrodes. The discharged electrode and the potassium hydroxide electrolyte are sent to a recycling facility. In a way, the zinc-air battery is analogous to a fuel cell, with the fuel being the zinc metal. A module of zinc air batteries tested in German Mercedes Benz postal vans had a specific energy of 200 Wh/kg, but only a modest specific power of 100 W/kg at 80% depth-of-discharge (see Sections 3.4 and 3.5 for definitions of depth-of-discharge and specific power). With present-day technology, the range of zinc-air batteries can be between 300 to 600 km between recharges.

Other metal-air systems have been investigated, but the work has been discontinued due to severe drawbacks in the technologies. These batteries include iron-air and aluminum-air batteries, in which iron and aluminum are, respectively, used as the mechanically recyclable negative electrode.

The practical metal-air batteries have two attractive features: the positive electrode can be optimized for discharge characteristics, because the battery is recharged outside of the battery; and the recharging time is rapid, with a suitable infrastructure.

3.3.6 SODIUM-SULFUR BATTERY

Sodium, similar to lithium, has a high electrochemical reduction potential (2.71 V) and low atomic mass (23.0), making it an attractive negative electrode element for batteries. Moreover, sodium is abundant in nature and available at a low cost. Sulfur, which is a possible choice for the positive electrode, is also a readily available and low-cost material. The use of aqueous electrolytes is not possible due to the highly reactive nature of sodium, and because the natures of solid polymers like those used for lithium batteries are not known. The solution of electrolyte came from the discovery of beta-alumina by scientists at Ford Motor Company in 1966. Beta-alumina is a sodium aluminum oxide with a complex crystal structure.

Despite several attractive features of NaS batteries, there are several practical limitations. The cell operating temperature in NaS batteries is around 300°C, which requires adequate insulation as well as a thermal control unit. This requirement forces a certain minimum size of the battery, limiting the development of the battery for only EVs, a market that is not yet established. Another disadvantage of NaS batteries is the absence of an overcharge mechanism. At the top-of-charge, one or more cells can develop a high resistance, which pulls down the entire voltage of the series-connected battery cells. Yet another major concern is safety, because the chemical reaction between molten sodium and sulfur can cause excessive heat or explosion in the case of an accident. Safety issues were addressed through efficient design, and manufactured NaS batteries have been shown to be safe.

Practical limitations and manufacturing difficulties of NaS batteries have led to the discontinuation of its development programs, especially when the simpler concept of sodium-metal-chloride batteries was developed.

3.3.7 SODIUM-METAL-CHLORIDE BATTERY

The sodium-metal-chloride battery is a derivative of the sodium-sulfur battery with intrinsic provisions of overcharge and overdischarge. The construction is similar to that of the NaS battery, but the positive sulfur electrode is replaced by nickel chloride ($NiCl_2$) or a mixture of nickel chloride and ferrous chloride ($FeCl_2$). The negative electrode and the electrolyte are the same as in a NaS battery. The schematic diagram of a $NaNiCl_2$ cell is shown in Figure 3.7. In order to provide good ionic contact between the positive electrode and the electrolyte, both of which are solids, a second electrolyte of sodium chloraluminate ($NaAlCl_4$) is introduced in a layer between $NiCl_2$ and beta-alumina. The $NaAlCl_4$ electrolyte is a vital component of the battery, although it reduces the specific energy of the battery by about 10%.[3] The operating temperature is again high, similar to that of the NaS battery. The basic cell reactions for the nickel chloride and ferrous chloride positive electrodes are as follows:

$$NiCl_2 + 2Na \quad \underset{Charge}{\overset{Discharge}{\rightleftarrows}} \quad Ni + 2NaCl \quad (2.58V)$$

$$FeCl_2 + 2Na \quad \underset{Charge}{\overset{Discharge}{\rightleftarrows}} \quad Fe + 2NaCl \quad (2.35V)$$

The cells in a sodium metal chloride battery are assembled in a discharged state. The positive electrode is prefabricated from a mixture of Ni or Fe powder and NaCl

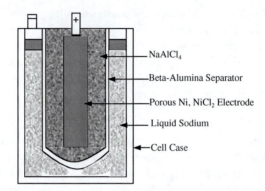

FIGURE 3.7 A sodium-nickel-chloride cell.

(common salt). On charging after assembly, the positive electrode compartment is formed of the respective metal, and the negative electrode compartment is formed of sodium. This procedure has two significant advantages: pure sodium is manufactured in situ through diffusion in beta-alumina, and the raw materials for the battery (common salt and metal powder) are inexpensive. Although iron is cheaper than nickel, the latter is more attractive as the metallic component because of fewer complications and a wider operating temperature range.

Sodium chloride batteries are commonly known as ZEBRA batteries, which originally resulted from a research collaboration between scientists from the United Kingdom and South Africa in the early 1980s. ZEBRA batteries have been shown to be safe under all conditions of use. They have high potential for being used as batteries for EVs and HEVs. There are several test programs utilizing the ZEBRA batteries.

3.4 BATTERY PARAMETERS

3.4.1 Battery Capacity

The amount of free charge generated by the active material at the negative electrode and consumed by the positive electrode is called the *battery capacity*. The capacity is measured in Ah (1 Ah = 3600 C, or coulomb, where 1 C is the charge transferred in 1 s by 1 A current in the MKS unit of charge).

The theoretical capacity of a battery (in C) is:

$$Q_T = xnF$$

where x is the number of moles of limiting reactant associated with complete discharge of the battery, n is the number of electrons produced by the negative electrode discharge reaction, and $F = Le_0$. L is the number of molecules or atoms in a mole (known as the Avogadro constant), and e_0 is the electron charge. F is the Faraday constant. The values for the constants are:

$$L = 6.022 \times 10^{23}$$

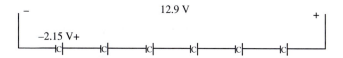

FIGURE 3.8 Battery cells connected in series.

and

$$e_0 = 1.601 \times 10^{-19} \, C$$

$$\Rightarrow F = 96412.2 \, C/mol$$

The theoretical capacity in Ah is:

$$Q_T = 0.278 \, F \frac{m_R n}{M_M} \tag{3.1}$$

where m_R is the mass of limiting reactant (in kg), and M_M is the molar mass of limiting reactant (in g/mol).

The cells in a battery are typically connected in series (Figure 3.8), and the capacity of the battery is dictated by the smallest cell capacity. Therefore, $Q_{Tbattery} = Q_{Tcell}$.

3.4.2 DISCHARGE RATE

The *discharge rate* is the current at which a battery is discharged. The rate is expressed as Q/h rate, where Q is rated battery capacity, and h is discharge time in hours. For a battery that has a capacity of Q_T Ah and that is discharged over Δt, the discharge rate is $Q_T/\Delta t$.

For example, let the capacity of a battery be 1 Q = 100 Ah. (1 Q usually denotes the rated capacity of the battery.) Therefore,

$$Q/5 \text{ rate is } \frac{100 \text{ Ahr}}{5 \text{ hr}} = 20 \, A$$

and

$$2Q \text{ rate is } Q/0.5 = \frac{100 \text{ Ahr}}{0.5 \text{ hr}} = 200 \, A$$

3.4.3 STATE OF CHARGE

The *state of charge* (*SoC*) is the present capacity of the battery. It is the amount of capacity that remains after discharge from a top-of-charge condition. The battery

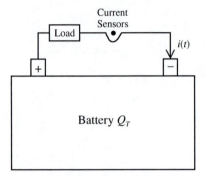

FIGURE 3.9 Battery *SoC* measurement.

SoC measurement circuit is shown in Figure 3.9. The current is the rate of change of charge given by

$$i(t) = \frac{dq}{dt}$$

where q is the charge moving through the circuit. The instantaneous theoretical state of charge $SoC_T(t)$ is the amount of equivalent positive charge on the positive electrode. If the state of charge is Q_T at the initial time t_o, then $SOC_T(t_o) = Q_T$. For a time interval dt,

$$dSoC_T = -dq$$
$$= -i(t)dt$$

Integrating from the initial time t_o to the final time t, the expression for instantaneous state of charge is obtained as follows:

$$SoC_T(t) = Q_T - \int_o^t i(\tau)d\tau \qquad (3.2)$$

3.4.4 State of Discharge

The *state of discharge* (*SoD*) is a measure of the charge that has been drawn from a battery. Mathematically, state of discharge is given as follows:

$$SoD_T(t) = \Delta q = \int_o^t i(\tau)d\tau$$
$$\Rightarrow SoC_T(t) = Q_T - SoD_T(t) \qquad (3.3)$$

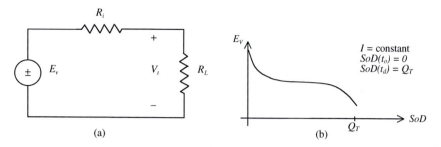

FIGURE 3.10 (a) Steady-state battery equivalent circuit. (b) Battery open circuit voltage characteristics.

3.4.5 DEPTH OF DISCHARGE

The *depth of discharge (DoD)* is the percentage of battery capacity (rated capacity) to which a battery is discharged. The depth of discharge is given by

$$DoD(t) = \frac{Q_T - SoC_T(t)}{Q_T} \times 100\%$$

$$= \frac{\int_o^t i(\tau)\,d\tau}{Q_T} \times 100\% \tag{3.4}$$

The withdrawal of at least 80% of battery (rated) capacity is referred to as deep discharge.

3.5 TECHNICAL CHARACTERISTICS

The battery in its simplest form can be represented by an internal voltage E_v and a series resistance R_i, as shown in Figure 3.10a. The internal voltage appears at the battery terminals as open circuit voltage when there is no load connected to the battery. The internal voltage or the open circuit voltage depends on the state of charge of the battery, temperature, and past discharge/charge history (memory effects), among other factors. The open circuit voltage characteristics are shown in Figure 3.10b. As the battery is gradually discharged, the internal voltage decreases, while the internal resistance increases. The open circuit voltage characteristics have a fairly extended plateau of linear characteristics, with a slope close to zero. The open circuit voltage is a good indicator of the state of discharge. Once the battery reaches 100% *DoD*, the open circuit voltage decreases sharply with more discharge.

The battery terminal voltage (Figure 3.11) is the voltage available at the terminals when a load is connected to the battery. The terminal voltage is at its full charge voltage V_{FC} when the battery *DoD* is 0%. In the case of a lead-acid battery, it means that there is no more $PbSO_4$ available to react with H_2O to produce active material. V_{cut} is the battery cut-off voltage, where discharge of battery must be terminated.

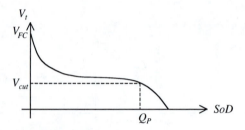

FIGURE 3.11 Battery terminal voltage.

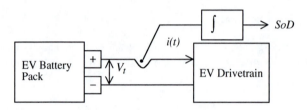

FIGURE 3.12 Battery *SoD* measurement.

In order to predict the range of an EV, the *SoC* or *DoD* can be used. However, the question is which one will be more accurate? The *SoC* and *DoD* are related as

$$SoC(t) = Q_T - SoD(t) \text{(in Ah)}$$

$$DoD = \frac{SoD}{Q_T}$$

The reliability of *SoC* depends on reliability of Q_T, which is a function of discharge current and temperature, among other things. It will be difficult to use *SoC* for general discharge currents, because it is hard to predict Q_T. On the other hand, the *DoD* can be expressed more accurately, because it is expressed as a fraction of Q_T, and it is easier to measure *SoD* (Figure 3.12).

3.5.1 PRACTICAL CAPACITY

The *practical capacity* Q_P of a battery is always much lower compared to the theoretical capacity Q_T, due to practical limitations. The practical capacity of a battery is given as

$$Q_P = \int_{t_o}^{t_{cut}} i(t) dt \tag{3.5}$$

where t_o is the time at which the battery is fully charged [$SoD(t_o) = 0$], and t_{cut} is the time at which the battery terminal voltage is at V_{cut}. Therefore, $V_t(t_{cut}) = V_{cut}$.

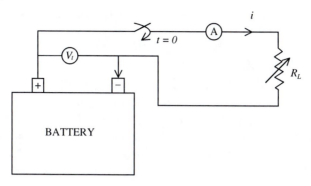

FIGURE 3.13 Battery capacity measurement.

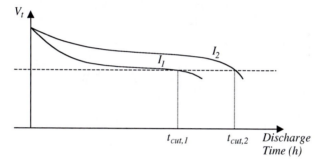

FIGURE 3.14 Constant current discharge curves.

3.5.1.1 Capacity Redefined

The practical capacity of a battery is defined in the industry by a convenient and approximate approach of Ah instead of coulomb under constant discharge current characteristics. Let us consider the experiment shown in Figure 3.13, where the battery is discharged at constant current starting from time $t = 0$. Current is maintained constant by varying the load resistance R_L until the terminal voltage reaches V_{cut}. The constant current discharge characteristics are shown qualitatively at two different current levels in Figure 3.14. The following data are obtained from the experiment:

$$I = 80\,\text{A: Capacity } Q_{80\,\text{A}} = (80\,\text{A})t_{cut} = 80 \times 1.8 = 144\,\text{Ah}$$

$$I = 50\,\text{A: Capacity } Q_{50\,\text{A}} = (50\,\text{A})t_{cut} = 50 \times 3.1 = 155\,\text{Ah}$$

$$I = 30\,\text{A: Capacity } Q_{30\,\text{A}} = (30\,\text{A})t_{cut} = 30 \times 5.7 = 171\,\text{Ah}$$

The results show that the capacity depends on the magnitude of discharge current. The smaller the magnitude of the discharge current, the higher the capacity of the battery. To be accurate, when the capacity of a battery is stated, the constant discharge current magnitude must also be specified.

3.5.1.2 Battery Energy

The energy of a battery is measured in terms of capacity and discharge voltage. To calculate energy, the capacity of the battery must be expressed in coulombs. A measurement of 1 Ah is equivalent to 3600 C, while 1 V refers to 1 J (J for joule) of work required to move 1 C charge from the negative to positive electrode. Therefore, the stored electrical potential energy in a 12 V, 100 Ah battery is $(12)(3.6 \times 10^5)$J = 4.32 MJ. In general, the theoretical stored energy is:

$$E_T = V_{bat}Q_T$$

where V_{bat} is the nominal no load terminal voltage, and Q_T is the theoretical capacity in C. Therefore, using Equation 3.1, we have

$$E_T = \left[\frac{1000\,Fn}{M_M} m_R \right] V_{bat} = 9.65 \times 10^7 \frac{nm_R}{M_M} V_{bat} \tag{3.6}$$

The practical available energy is:

$$E_P = \int_{t_o}^{t_{cut}} vi\,dt \ \text{Wh} \tag{3.7}$$

where t_o is the time at which the battery is fully charged, t_{cut} is the time at which battery terminal voltage is at V_{cut}, v is the battery terminal voltage, and i is the battery discharge current. E_P is dependent on the manner in which the battery is discharged.

3.5.1.3 Constant Current Discharge

The battery terminal voltage characteristic is shown again in Figure 3.15, indicating the midpoint voltage (MPV) at $t = 1/2\,t_{cut}$. The extended plateau of the midpoint voltage can be represented by the straight-line equation $V_t = mt + b$, where m and b are the constants of the equation. We will replace the nonlinear terminal voltage characteristic of the battery by the extended plateau straight-line equation for simplicity. The energy of a battery with constant current discharge is

$$E_I = \int_o^{t_{cut}} V_t I\,dt = I \int_o^{t_{cut}} V_t\,dt$$

Let the average battery terminal voltage over discharge interval 0 to t_{cut} be $<v_t>$. Therefore,

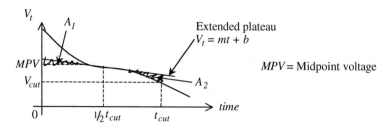

FIGURE 3.15 Midpoint voltage.

$$< v_t >= \frac{1}{t_{cut}} \int_o^{t_{cut}} V_t \, dt = \frac{1}{t_{cut}} \left[A_1 + \int_o^{t_{cut}} (mt+b) \, dt - A_2 \right]$$

where A_1 and A_2 are the areas indicated in Figure 3.15. We can assume that the areas A_1 and A_2 are approximately equal, $A_1 \cong A_2$. However, $m(1/2 \, t_{cut}) + b = MPV$ (MPV is midpoint voltage). This gives

$$< v_t >= MPV \Rightarrow \int_o^{t_{cut}} V_t \, dt = t_{cut} \, MPV$$

Substituting,

$$E_I = I \cdot t_{cut} \cdot MPV = Q_I \cdot MPV \tag{3.8}$$

An empirical relation often used to describe battery characteristics is *Peukert's equation*. Peukert's equation relating constant current with t_{cut} is as follows:

$$t_{cut} = \frac{\lambda}{I^n} \tag{3.9}$$

where λ and n are constants. Substituting Peukert's equation in the energy equation,

$$E_I \cong \lambda I^{1-n} \, MPV$$

3.5.1.4 Specific Energy

The *specific energy* of a battery is given by

$$SE = \frac{\text{Discharge Energy}}{\text{Total Battery Mass}} = \frac{E}{M_B}$$

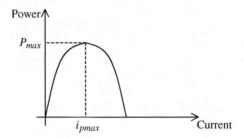

FIGURE 3.16 Battery power characteristics.

The unit for specific energy is Wh/kg. The theoretical specific energy of a battery is

$$SE_T = 9.65 \times 10^7 \times \frac{nV_{bat}}{M_M} \frac{m_R}{M_B} \qquad (3.10)$$

If the mass of the battery M_B is proportional to the mass of the limiting reactant of the battery m_R, then SE_T is independent of mass. The specific energy of a lead-acid battery is 35 to 50 Wh/kg at Q/3 rate. Because E_p varies with discharge rate, the practical specific energy SE_P is also variable.

The term energy density is also used in the literature to quantify the quality of a battery or other energy source. Energy density refers to the energy per unit volume of a battery. The unit for energy density is Wh/liter.

3.5.2 BATTERY POWER

Battery power characteristics are illustrated in Figure 3.16. The instantaneous battery terminal power is:

$$p(t) = V_t i \qquad (3.11)$$

where V_t is the battery terminal voltage, and i is the battery discharge current. Using Kirchoff's voltage law for the battery equivalent circuit of Figure 3.10a,

$$V_t = E_v - R_i i \qquad (3.12)$$

Substituting Equation 3.12 into Equation 3.11 yields:

$$p(t) = E_v i - R_i i^2 \qquad (3.13)$$

Maximum power output is needed from the battery in fast discharge conditions, which occur when the electric motor is heavily loaded. Acceleration on a slope is such a condition, when the motor draws a lot of current to deliver maximum power required for traction. Using the maximum power transfer theorem in electric circuits, the battery can deliver maximum power to a DC load when the load impedance matches the battery internal impedance. The maximum power is:

$$P_{max} = \frac{E_v^2}{4R_t}$$

Because E_v and R_i vary with the state of charge, P_{max} also varies accordingly.

The performance of batteries to meet acceleration and hill climbing requirements can be evaluated with the help of rated power specifications, which are based on the ability of the battery to dissipate heat. *Rated continuous power* is the maximum power that the battery can deliver over prolonged discharge intervals without damage to the battery. These do not necessarily correspond to P_{max} on the p–i curve of battery characteristics. The *rated instantaneous power* is the maximum power that the battery can deliver over a short discharge interval without damage to the battery.

3.5.2.1 Specific Power

The specific power of a battery is

$$SP = \frac{P}{M_B} \left(\text{units: Wh/kg}\right) \tag{3.14}$$

where P is the power delivered by the battery, and M_B is the mass of the battery. Typically, a lead-acid battery's maximum specific power is 280 W/kg (which corresponds to P_{max}) at $DoD = 80\%$. Similar to specific energy and energy density, power density refers to the power of the battery per unit volume, with units of W/liter.

3.5.2.2 Battery Pack Design

Batteries can be configured in series or in parallel configuration or in a combination thereof. Design depends on output voltage and discharge requirements. The series connection yields the required voltage, while the parallel connection delivers the desired capacity for the battery pack for minimum run time before recharging. The battery pack also includes electronics, which are typically located outside the battery pack. The electronic circuit of a multilevel battery pack controls charging and ensures reliability and protection against overcharge, overdischarge, short circuiting, and thermal abuse.

3.5.3 RAGONE PLOTS

In lead-acid and other batteries, there is a decrease in charge capacity (excluding voltage effects) with increasing currents. This is often referred to as the *Ragone* relationship and is described by *Ragone plots*. Ragone plots are usually obtained from constant power discharge tests or constant current discharge plots. Consider the experiment of Figure 3.13, but this time, the current i is adjusted by varying R_L

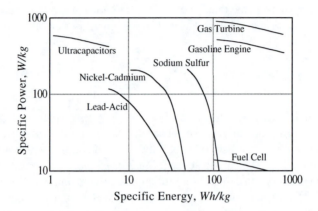

FIGURE 3.17 Specific power vs. specific energy (Ragone plots) of batteries, gasoline engine, and fuel cell.

such that the power output at the battery terminals is kept constant. The experiment stops when the battery terminal voltage reaches the cut-off voltage, i.e., $V_t = V_{cut}$. We assume that the battery is fully charged at $t = 0$. The experiment is performed at several power levels, and the following data are recorded: power $p(t) = V_t i = P$, time to cut-off voltage t_{cut}, and practical energy $E_p = P t_{cut}$. The plot of SP vs. SE on log–log scale is known as the Ragone plot. The Ragone plots of some common batteries are shown in Figure 3.17.

To a first-order approximation, we can use a linear Ragone plot (on a log–log scale) according to the following relationship between specific power and specific energy:

$$(SP)^n (SE) = \lambda \tag{3.15}$$

where n and λ are curve-fitting constants. The above is an alternative approach of using Peukert's equation to describe battery characteristics. The Ragone plots of several batteries, along with alternative energy sources and internal combustion (IC) engines, are given in Figure 3.17 to give an idea about the relative power and energy capacities of these different units.

EXERCISE 3.1

The data given in Table 3.3 are collected from an experiment on a battery with mass 15 kg. The data are used to draw the Ragone plot shown in Figure 3.18. Using the data points (8,110) and (67.5,10), calculate the constants of Peukert's equation, n and λ.

Solution

$$n = 0.8894 \quad \text{and} \quad \lambda = 523.24 \text{ Ah}$$

TABLE 3.3
Data from Constant Power Discharge Test

P (W) (Measured)	t_{cut} (h) (Measured)	E_p (Wh) (Calculated)	SP (W/kg) (Calculated)	SE (Wh/kg) (Calculated)
150	6.75	(150)(6.75) = 1013	150/15 = 10	1013/15 = 67.5
450	0.85	381	30	25.4
900	0.23	206	60	13.7
1650	0.073	120	110	8

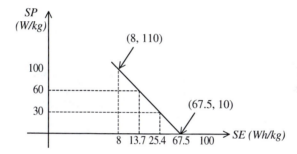

FIGURE 3.18 Ragone plot for Exercise 3.1.

3.6 TARGETS AND PROPERTIES OF BATTERIES

The push for zero-emission vehicles led to numerous research and development initiatives in the United States, Europe, and Japan. The California legislative mandates in the early 1990s led to the formation of the U.S. Advanced Battery Consortium (USABC) to oversee the development of power sources for EVs. The USABC established objectives focusing on battery development for mid-term (1995 to 1998) and long-term criteria. The purpose of the mid-term criteria was to develop batteries with a reasonable goal, while the long-term criteria were set to develop batteries for EVs, which would be directly competitive with IC engine vehicles (ICEVs). At the advent of the 21st century and following the developments in the 1990s, intermediate commercialization criteria were developed. The major objectives for the three criteria are summarized in Table 3.4.

The two most developed battery technologies of today are lead-acid and nickel-cadmium batteries. However, these batteries will not be suitable for EVs, because the former store too little energy, while the latter have cost and toxicity problems. The future of the other batteries is difficult to predict, because these are mostly prototypes, where system design and performance data are not always available. The status of the promising batteries described in the previous section is summarized in Table 3.5 from information obtained from recent literature.

TABLE 3.4
USABC Objectives for EV Battery Packs

Parameter	Mid-Term	Commercialization	Long-Term
Specific energy (Wh/kg) (C/3 discharge rate)	80–100	150	200
Energy density (Wh/liter) (C/3 discharge rate)	135	230	300
Specific power (W/kg) (80% *DoD* per 30 s)	150–200	300	400
Specific power (W/kg), Regen. (20% *DoD* per 10 s)	75	150	200
Power density (W/liter)	250	460	600
Recharge time, h (20% → 100% *SoC*)	<6	4–6	3–6
Fast recharge time, min	<15	<30	<15
Calendar life, years	5	10	10
Life, cycles	600 @ 80% *DoD*	1000 @ 80% *DoD* 1600 @ 50% *DoD* 2670 @ 30% *DoD*	1000 @ 80% *DoD*
Lifetime urban range, miles	100,000	100,000	100,000
Operating environment, °C	–30 to +65	–40 to +50	–40 to +85
Cost, US$/kWh	<150	<150	<100
Efficiency, %	75	80	80

TABLE 3.5
Properties of EV and HEV Batteries

Battery Type	Specific Energy, Wh/kg	Specific Power, W/kg	Energy Efficiency, %	Cycle Life	Estimated Cost, US$/kWh
Lead-acid	35–50	150–400	80	500–1000	100–150
Nickel-cadmium	30–50	100–150	75	1000–2000	250–350
Nickel-metal-hydride	60–80	200–300	70	1000–2000	200–350
Aluminum-air	200–300	100	<50	Not available	Not available
Zinc-air	100–220	30–80	60	500	90–120
Sodium-sulfur	150–240	230	85	1000	200–350
Sodium-nickel-chloride	90–120	130–160	80	1000	250–350
Lithium-polymer	150–200	350	Not available	1000	150
Lithium-ion	80–130	200–300	>95	1000	200

3.7 BATTERY MODELING

Peukert's equation is a widely accepted empirical relation among capacity (Q), discharge current (I), and time (t) or among specific power (SP), specific energy (SE), and time (t). Peukert's equation is used in developing a *fractional depletion model* (FDM) of batteries. The FDM of a battery can be used to predict the range of an EV. The FDM can be developed using the constant current discharge approach or the power density approach associated with the two forms of Peukert's equation.

3.7.1 CONSTANT CURRENT DISCHARGE APPROACH

Consider the constant current discharge experiment of Figure 3.13. The battery is discharged under constant current condition from 100% capacity until cut-off voltage is reached. The load resistance R_L is varied to change the constant current level and also to maintain the current constant for each experiment. The I vs. t_{cut} data are used to fit Peukert's equation (Equation 3.9) with constant current:

$$I^n * t_{cut} = \lambda \tag{3.16}$$

where I is the constant discharge current; and λ and n are curve-fitting constants, with $n \rightarrow 1$ for small currents and $n \rightarrow 2$ for large currents.

EXAMPLE 3.1

Find the curve-fitting constants n and λ for Peukert's equation for the two measurements available from a constant current discharge experiment of a battery:

(i) $(t_1, I_1) = (10,18)$
(ii) $(t_2, I_2) = (1,110)$

Solution

Equation 3.16 is the Peukert's empirical formula using the constant current discharge approach. Taking logarithm of both sides of Equation 3.16:

$$Log_{10}\left(I^n * t_{cut}\right) = Log_{10}(\lambda)$$

$$\Rightarrow Log_{10}(I) = \frac{1}{n}Log_{10}(t_{cut}) + \frac{1}{n}Log_{10}(\lambda)$$

Comparing with the equation for a straight line, $y = mx + b$, I vs. t_{cut} curve is linear on a log–log plot, as shown in Figure 3.19. The slope of the straight line is

$$m = \frac{\Delta y}{\Delta x} = \frac{\log(I_1) - \log(I_2)}{\log(t_1) - \log(t_2)} = \frac{\log\left(\dfrac{I_1}{I_2}\right)}{\log\left(\dfrac{t_1}{t_2}\right)}$$

Therefore,

$$n = -\frac{\log\left(\dfrac{t_1}{t_2}\right)}{\log\left(\dfrac{I_1}{I_2}\right)}$$

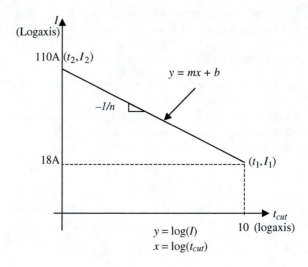

FIGURE 3.19 Plot of Peukert's equation using constant current discharge.

For the graph shown,

$$n = \frac{-1}{18/110} = 1.27 \qquad \left[\because t_1 = 10t_2 \right]$$

The other constant can now be calculated from Peukert's equation:

$$\lambda = 110^{1.27} * 1 = 3.91.4 \text{ Ah} \quad \text{or}$$

$$\lambda = 18^{1.27} * 10 = 392.8 \text{ Ah}$$

3.7.1.1 Fractional Depletion Model

Using Peukert's equation, we can establish the relationship between Q and I. The practical capacity of a battery is

$$Q = I * t_{cut}$$

$$\Rightarrow t_{cut} = \frac{Q}{I}$$

Substituting into Peukert's equation:

$$I^n \left(\frac{Q}{I} \right) = \lambda$$

$$\Rightarrow Q = \frac{\lambda}{I^{n-1}}$$

Because $0 < n - 1 < 1$, for $I > 1$, Q decreases as I increases.

From Section 3.4, we know that

$$SoD = \int i(\tau)\,d\tau$$

and

$$DoD = \frac{SoD}{Q(i)}$$

SoD is the amount of charge that the battery generates to the circuit. Assume that at $t = t_0$, the battery is fully charged. Let us consider a small interval of time dt. Therefore,

$$d(DoD) = \frac{d(SoD)}{Q(i)}$$

where

$$d(SoD) = i(t)\,dt$$

We know that $Q = \lambda/I^{n-1}$ for constant current discharge. Let $Q = \lambda/i^{n-1}$ for time varying current as well, for lack of anything better.

Therefore,

$$d(DoD) = \frac{idt}{\lambda/i^{n-1}} = \frac{i^{n}}{\lambda}\,dt$$

Integrating, we obtain,

$$\int_{t_0}^{t} d(DoD) = \int_{t_0}^{t} \frac{i^{n}}{\lambda}\,dt$$

$$\Rightarrow DoD(t) - DoD(t_0) = \int_{t_0}^{t} \frac{i^{n}}{\lambda}\,dt$$

$DoD(t_0) = 0$ if the battery is fully charged at $t = t_0$.

The *fractional depletion model* (FDM) is thus obtained as

$$DoD(t) = \left[\int_{t_0}^{t} \frac{i^{n}}{\lambda}\,dt \right] * 100\% \tag{3.17}$$

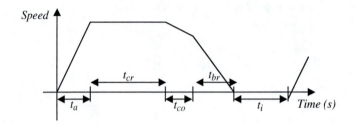

FIGURE 3.20 SAE J227a standard driving cycle.

The FDM based on current discharge requires knowledge of the discharge current $i(t)$. Therefore, this model to predict the EV range should be used when $i(t)$ is known.

3.7.2 STANDARD DRIVING CYCLES

The standard J227a driving cycle recommended by the Society of Automotive Engineers (SAE) is routinely used to evaluate the performance of EVs and energy sources. The SAE J227a has three schedules designed to simulate the typical driving patterns of fixed-route urban, variable-route urban, and variable-route suburban travels. These three patterns are the SAE J227a driving schedules B, C, and D, respectively. Each schedule has five segments in the total driving period:

1. Acceleration time t_a to reach the maximum velocity from start-up
2. Cruise time t_{cr} at a constant speed
3. Coast time t_{co} when no energy is drawn from the source
4. Brake time t_{br} to bring the vehicle to stop
5. Idle time t_i prior to the completion of the period

The driving cycle for J227a is shown in Figure 3.20, with the recommended times for each of the schedules given in Table 3.6. The figure drawn is slightly modified from the pattern recommended by the SAE. The J227a procedures specify only the cruise velocity and the time of transition from one mode to the other. The velocity profile at segments other than the cruising part is not fixed, and hence, the distance traversed during these other periods is also variable. In reality, the distances would depend on the design of the vehicle under consideration. For simplicity, straight-line approximations have been assumed in these schedules in this book.

EXAMPLE 3.2

The constant current discharge characteristics of the battery pack used in an EV are as follows:

$$\ln I = 4.787 - 0.74 \ln t_{cut} - 0.0482 \left(\ln t_{cut} \right)^2$$

TABLE 3.6
SAE J227a Standard Driving Schedules

Test Parameter	SAE J227a Schedules		
	B	C	D
Maximum speed, km/h	32	48	72
Acceleration time, t_a s	19	18	28
Cruise time, t_{cr} s	15	20	50
Coast time, t_{co} s	4	8	10
Brake time, t_{br} s	5	9	9
Idle time, t_i s	25	25	25
Total time, s	72	80	122
Approximate number of cycles per mile	4–5	3	1

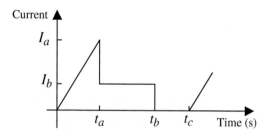

FIGURE 3.21 Pattern of current drawn from the battery.

TABLE 3.7
Current Data for the Driving Schedules

Schedule J227a	I_a (A)	I_b (A)
B	100	35
C	216	54.6
D	375	88.7

The current drawn from the battery during test drives of the EV for the SAE schedule J227a has the profile shown in Figure 3.21. The current magnitudes for the three SAE schedules are given in Table 3.7.

Find the range of the EV for each of the three schedules.

Solution

Apply the FDM (Equation 3.17) to find the number of driving cycles for $DoD = 100\%$. From FDM:

$$1 = \int_{t_0}^{t_{100\%}} \frac{i_n}{\lambda} \, dt \qquad (3.18)$$

First, we need to determine λ and n from the given battery characteristics:

$$\frac{-1}{n} = -0.74 \Rightarrow n = 1.35$$

$$\frac{1}{n} \ln(\lambda) = 4.787 \Rightarrow \lambda = 645 * 3600 \ A - \sec$$

Therefore,

$$1 = \int_0^{t_{100\%}} \frac{i^{1.35}}{645 * 3600} \, dt$$

For Schedule B, the fraction depleted over one cycle is as follows:

$$DoD \text{ for 1 cycle} \Rightarrow f_{cyc} = \int_0^{72} \frac{i^{1.35}}{645 * 3600} \, dt$$

$$\Rightarrow f_{cyc} = 4.31 * 10^{-7} \left[\int_0^{19} \left(\frac{100t}{19} \right) dt + \int_{19}^{38} (35)^{1.35} \, dt \right]$$

$$= 4.31 * 10^{-7} \left[9.41 \left(\frac{1}{2.35} \right) 119^{2.35} + 121.5(38 - 19) \right]$$

$$\Rightarrow f_{cyc} = 2.74 * 10^{-3}$$

Let N = the number of cycles required for 100% DoD, $DoD = 1$:

$$\therefore 1 = N * f_{cyc} \Rightarrow N = \frac{1}{f_{cyc}}$$

$$\therefore N = \frac{1}{2.74 * 10^{-3}} = 365 \text{ cycles}$$

From Table 3.6, the EV goes 1 mi in about four cycles for Schedule B.

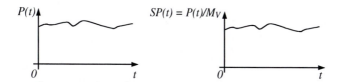

FIGURE 3.22 Power and specific power.

Therefore,

$$\text{EV Range } = \frac{365}{4} = 91 \text{ miles for } \textbf{Schedule B}$$

$$\text{Measured } N = 369 \Rightarrow \text{Error } = 1.08\%$$

J227a Schedule C: From FDM, $N = 152$; EV range $= 152/3 = 51$ mi. (Measured, $N = 184 \Rightarrow$ Error $= 17.4\%$.)

J227a Schedule D: From FDM, $N = 41$; EV range $= 41/1 = 41$ mi. (Measured, $N = 49 \Rightarrow$ Error $= 16.3\%$.)

3.7.3 POWER DENSITY APPROACH

Given a battery terminal power profile $p(t)$, the specific power $SP(t)$ profile can be obtained by dividing the power profile $p(t)$ by the total vehicle mass M_V. The battery is assumed to be fully charged at $t = 0$. Let $f_y(t)$ be equal to the fraction of available energy provided by the battery from 0 to t, where $f_r(0) = 0$, because $SoD(0) = 0$. Now, consider the time interval dt over which a fraction of available energy df_r is provided by the battery:

$$df_r = \frac{dE}{E_{avail}} = \frac{\dfrac{dE}{M_v}}{\dfrac{dE_{avail}}{M_v}} = \frac{d(SE)}{SE_{avail}}$$

If dE is the energy provided by the battery to the electrical circuit over dt, and E_{avail} is the total available energy, then

$$dE = pdt$$

Now E_{avail} is a function of instantaneous power, and we know that

$$d(SE) = (SP)dt$$

Therefore,

$$SE_{avail} = f(SP)$$

We will use Peukert's equation to relate specific power and specific energy as follows:

$$(SP)^n * SE_{avail} = \lambda$$

Therefore,

$$df_r = \frac{SP}{\dfrac{\lambda}{(SP)^n}} dt = \frac{(SP)^{n+1}}{\lambda} dt$$

Integrating,

$$\int_{f_r(0)}^{f_r(t)} df_r = \int_0^t \frac{(SP)^{n+1}}{\lambda} d\tau$$

$$\Rightarrow f_r(t) = \int_0^t \frac{(SP)^{n+1}}{\lambda} d\tau$$

(3.19)

Equation 3.19 is the FDM using the power density approach. If t is the time at which $x\%$ of available energy has been used, then

$$\frac{x}{100} = \int_0^t \frac{(SP)^{n+1}}{\lambda} d\tau$$

Note that

$$1 = \int_0^{t_{100\%}} \frac{(SP)^{n+1}}{\lambda} d\tau$$

At $t_{100\%}$, 100%, all of the available energy has been used by the system.

REFERENCES

1. Rand, D.A.J., Woods, R., and Dell, R.M., *Batteries for Electric Vehicles*, John Wiley & Sons, New York, 1998.

2. Dhameja, S., *Electric Vehicle Battery Systems*, Newnes (Elsevier Science), Burlington, MA, 2002.
3. Dell, R.M. and Rand, D.A.J., *Understanding Batteries*, Royal Society of Chemistry, United Kingdom, 2001.

PROBLEMS

3.1

Estimate the weight of a 12 V, 100 Ah lead-acid battery by calculating the reactant masses participating in the overall chemical reaction. Assume that the mass of H_2O in the electrolyte is twice the mass of H_2SO_4. Neglect battery casing mass, electrode grid mass, separator mass, and current bus mass. (Note that $n = 2$ for Pb and PbO_2, and $n = 1$ for H_2SO_4.)

3.2

In the nickel-cadmium cell, nickel oxyhydroxide NiOOH is the active material in the charged positive plate. During discharge, it reduces to the lower valence state, nickel hydroxide $Ni(OH)_2$, by accepting electrons from the external circuit:

$$2NiOOH + 2H_2O + 2e^- \; \overset{\text{Discharge}}{\underset{\text{Charge}}{\rightleftarrows}} \; 2Ni(OH)_2 + 2OH^- \qquad (0.49 \text{ volts})$$

Cadmium metal is the active material in the charged negative plate. During discharge, it oxidizes to cadmium hydroxide $Cd(OH)_2$ and releases electrons to the external circuit:

$$Cd + 2OH^- \; \overset{\text{Discharge}}{\underset{\text{Charge}}{\rightleftarrows}} \; Cd(OH)_2 + 2e^- \qquad (0.809 \text{ volts})$$

The net reaction occurring in the potassium hydroxide (KOH) electrolyte is:

$$Cd + 2NiOOH + 2H_2O \; \overset{\text{Discharge}}{\underset{\text{Charge}}{\rightleftarrows}} \; 2Ni(OH)_2 + Cd(OH)_2^- \qquad (1.299 \text{ volts})$$

Estimate the weight of an 11.7 V, 100 Ah Ni-Cd battery. Neglect the mass of the KOH component of the electrolyte.

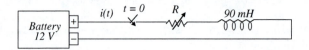

FIGURE P3.3

3.3

A 12 V battery is connected to a series RL load as shown in Figure P3.3. The battery has a rated capacity of 80 Ah. At $t = 0$, the switch is closed, and the battery begins to discharge.

(a) Calculate and plot the battery discharge current, $i(t)$, if the steady state discharge rate is $Q/2$. Neglect battery voltage drop.
(b) Calculate and plot $SoD(t)$ in Ah for $0 < t < 2$ h.
(c) Calculate and plot $SoC(t)$, assuming that at $t = 0$ the battery is charged to rated capacity. Assume also that the rated capacity is the practical capacity.
(d) Calculate the time corresponding to 80% *DoD*.

3.4

Given below are constant power discharge characteristics of a 12 V lead-acid battery:

SP (W/kg)	SE (Wh/kg)
10	67.5
110	8

The battery characteristics are to be expressed in terms of Peukert's equation, which has the following form:

$$(SP)^n (SE) = \lambda \qquad (n \text{ and } \lambda \text{ are curve fitting constants})$$

(a) Derive the constants n and λ, assuming a linear relationship between $\log(SP)$ and $\log(SE)$.
(b) Find the capacity Q_T of the battery if the theoretical energy density is $SE_T = 67.5$ Wh/kg, given a battery mass of 15 kg.

3.5

An EV battery pack consists of four parallel sets of six series-connected 12 V, 100 Ah lead-acid batteries. One steady state motoring (discharge) cycle of battery current is shown in Figure P3.5a. The steady-state regenerative braking (charge) cycle of the battery is shown in Figure P3.5b.

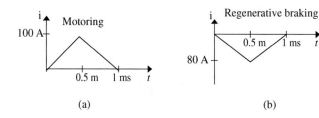

(a)   (b)

FIGURE P3.5

(a) Suppose no regenerative braking is employed. How much time does it take to reach 80% *DoD*?

(b) If regenerative braking is employed such that for every 50 motoring cycles there is one regenerative braking cycle, how much time does it take to reach 80% *DoD*?

(Note: In this problem, neglect variation of capacity with discharge rate. Assume that the practical capacity is equal to the rated capacity.)

3.6

Given a lead-acid battery having the following empirical characteristics:

$$(SP)^{.9}(SE) = 600 \text{ Ah}$$

where *SP* is specific power, and *SE* is specific energy. The EV parameters are as follows:

$$m = 700 \text{ kg}, M_B = 150 \text{ kg}, C_D = 0.2, A_F = 2 \text{ m}^2, C_0 = 0.009, \text{ and } C_1 = 0.$$

Also, take

$$\rho = 1.16 \text{ kg/m}^3, \text{ and } g = 9.81 \text{ m/s}^2$$

(a) Derive and plot $F_{TR}(t)$ vs. *t* (assume level road).

(b) Derive and plot $P_{TR}(t)$ vs. *t*.

(c) Calculate the EV range based on the SAE J227a Schedule B driving cycle using the power density approach of the FDM. The SAE J227a driving cycle and the current profile of the EV are given in Figures P3.6a and P3.6b. (Assume no regenerative braking.)

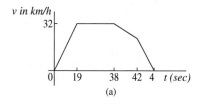

(a)

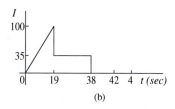

(b)

FIGURE P3.6

4 Alternative Energy Sources

The possible alternatives to batteries as portable energy sources that are being investigated today for electric vehicles (EVs) and hybrid electric vehicles (HEVs) and other applications are fuel cells and flywheels. Ultracapacitor technology has advanced tremendously in recent years, although it is unlikely to achieve specific energy levels high enough to serve as the sole energy source of a vehicle. However, ultracapacitors in conjunction with a battery or fuel cell have the possibility of being excellent portable energy sources with sufficient specific energy and specific power for the next generation of vehicles. These three alternative energy sources are covered in this chapter.

4.1 FUEL CELLS

A fuel cell is an electrochemical device that produces electricity by means of a chemical reaction, much like a battery. The major difference between batteries and fuel cells is that the latter can produce electricity as long as fuel is supplied, while batteries produce electricity from stored chemical energy and, hence, require frequent recharging.

The basic structure of a fuel cell (Figure 4.1) consists of an anode and a cathode, similar to a battery. The fuel supplied to the cell is hydrogen and oxygen. The concept of fuel cell is the opposite of electrolysis of water, where hydrogen and oxygen are combined to form electricity and water. The hydrogen fuel supplied to the fuel cell consists of two hydrogen atoms per molecule chemically bonded together in the form H_2. This molecule includes two separate nuclei, each containing one proton, while sharing two electrons. The fuel cell breaks apart these hydrogen molecules to produce electricity. The exact nature of accomplishing the task depends on the fuel cell type, although what remains the same for all fuel cells is that this reaction takes place at the anode. The hydrogen molecule breaks into four parts at the anode due to the chemical reaction, releasing hydrogen ions and electrons. A catalyst speeds the reaction, and an electrolyte allows the two hydrogen ions, which essentially are two single protons, to move to the cathode through the electrolyte placed between the two electrodes. The flow of electrons from the anode to the cathode through the external circuit is what produces electricity. For the overall cell reaction to complete, oxygen or air must be passed over the cathode. The cathode reaction takes place in two stages. First, the bond between the two oxygen atoms in the molecule breaks and then each ionized oxygen atom grabs two electrons coming from the anode through the external circuit to become negatively charged. The negatively charged

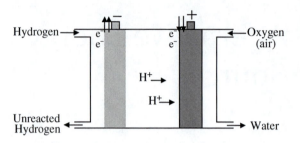

FIGURE 4.1 Basic fuel cell structure.

oxygen atoms are balanced by the positively charged hydrogen atoms at the cathode, and the combination produces H_2O commonly known as water. The chemical reaction taking place in a fuel cell is as follows:

$$\text{Anode: } H_2 \rightarrow 2H^+ + 2e^-$$

$$\text{Cathode: } 2e^- + 2H^+ + \frac{1}{2}(O_2) \rightarrow H_2O$$

$$\text{Cell: } H_2 + \frac{1}{2}O_2 \rightarrow H_2O$$

The fuel cell was first developed for space applications as an alternative power source. The source was first used in a moon buggy and is still used in NASA's space shuttles. There has been tremendous interest in fuel cells in recent years for applications in other areas, such as EVs and stationary power systems. The research sponsored by several U.S. research agencies and corporations has attempted to improve cell performance with two primary goals: a desire for higher power cells, which can be achieved through higher rates of reaction, and the desire for fuel cells that can internally reform hydrocarbons and are more tolerant of contaminants in the reactant streams. For this reason, the searches have concentrated on finding new materials for electrodes and electrolytes. There are several different types of fuel cells, each with strengths and weaknesses. Low operating temperature is desirable for vehicle applications, despite the fact that higher temperatures result in higher reaction rates. Rapid operation and cogeneration capabilities are desirable for stationary applications. Cogeneration refers to the capability to utilize the waste heat of a fuel cell to generate electricity using conventional means.

4.1.1 FUEL CELL CHARACTERISTICS

Theoretically, fuel cells operate isothermally, meaning that all free energy in a fuel cell chemical reaction should convert into electrical energy. The hydrogen "fuel" in the fuel cell does not burn as in IC engines, bypassing the thermal to mechanical conversion. Also, because the operation is isothermal, the efficiency of such direct electrochemical converters is not subject to the limitation of Carnot cycle efficiency imposed on heat engines. The fuel cell converts the Gibbs free energy of a chemical reaction into electrical energy in the form of electrons under isothermal conditions. The maximum electrical energy for a fuel cell operating at constant temperature and pressure is given by the change in Gibbs free energy:

$$W_{el} = -\Delta G = nFE \qquad (4.1)$$

where n is the number of electrons produced by the anode reaction; F is Faraday's constant, equal to 96412.2 C/mol; and E is the reversible potential. The Gibbs free energy change for the reaction $H_2(g) + (1/2)O_2(g) \rightarrow H_2O(l)$ at standard condition of 1 atmospheric pressure and 25°C is –236 kJ/mol or –118 MJ/kg. With $n = 2$, the maximum reversible potential under the same conditions is $E_0 = 1.23$ V, using Equation 4.1. The maximum reversible potential under actual operating conditions for the hydrogen-oxygen fuel cell is given by the Nernst equation, as follows:[1]

$$E = E_0 + \left(\frac{RT}{nF}\right) \ln\left[\frac{P_H \cdot P_O^{1/2}}{P_{H_2O}}\right] \qquad (4.2)$$

where T is the temperature in Kelvin; R is the gas constant; and P_H, P_O, and P_{H_2O} are the concentrations or partial pressures of the reactants and products.

The voltage-current output characteristic of a hydrogen-oxygen cell is illustrated in Figure 4.2. The higher potentials around 1 V per cell are theoretical predictions that are not achievable in a practical cell. The linear region where the reduction in cell potential is due to ohmic losses is where a practical fuel cell operates. The resistive components in the cell limit the practical achievable efficiency of a fuel cell. The working voltage of the cell falls with an increasing current drain, knowledge that is important in designing fuel-cell-powered EVs and hybrid vehicles. Because cell potential is small, several cells are stacked in series to achieve the desired voltage. The major advantage of fuel cells is lower sensitivity to scaling, which means that fuel cells in the kW range have similar overall system efficiencies up to the MW range.

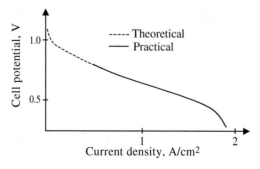

FIGURE 4.2 Voltage–current relationship of a hydrogen/oxygen cell.

4.1.2 FUEL CELL TYPES

The six major types of fuel cells are as follows: alkaline, proton exchange membrane, direct methanol, phosphoric acid, molten carbonate, and solid oxide. A short description of the relevant characteristics of each type in the context of vehicular and stationary applications is given below.[2,3]

4.1.2.1 Alkaline Fuel Cell (AFC)

In an alkaline fuel cell (AFC), an aqueous solution of potassium hydroxide (KOH) is used as the electrolyte. Compared to some other fuel cells where acidic electrolytes are used, the performance of the alkaline electrolyte is as good as the acid electrolytes, while being significantly less corrosive toward the electrodes. Alkaline fuel cells have been in actual use for a long time, delivering electrical efficiencies of up to 60%. They require pure hydrogen as fuel and operate at low temperatures (at 80°C); therefore, they are suitable for vehicle applications. Residual heat can be used for heating, but the cell temperature is not sufficiently high to generate steam that can be used for cogeneration.

4.1.2.2 Proton Exchange Membrane (PEM)

The proton exchange membrane (PEM) fuel cells use solid electrolytes and operate at low temperatures (around 80°C). Nafion is an example of solid polymer electrolyte. These fuel cells are also known as solid polymer membrane fuel cells. The electrical efficiency of PEM fuel cells is lower than that of the alkaline cells (about 40%). However, a rugged and simple construction makes these types of fuel cells suitable for vehicle applications. The PEM fuel cell and the AFC are currently being considered for vehicle applications. The advantage of PEM cells is that they can tolerate impurity in the fuel, as compared to pure hydrogen which is needed in alkaline fuel cells.

4.1.2.3 Direct Methanol Fuel Cell (DMFC)

The direct methanol fuel cell (DMFC) is a result of research on using methanol as the fuel that can be carried on-board a vehicle and reformed to supply hydrogen to the fuel cell. A DMFC works on the same principle as the PEM, except that the temperature is increased to the range of 90 to 120°C such that internal reformation of methanol into hydrogen is possible. The electrical efficiency of DMFC is quite low at about 30%. This type of fuel cell is still in the design stages, because the search for a good electrocatalyst to reform the methanol efficiently and to reduce oxygen in the presence of methanol is ongoing.

4.1.2.4 Phosphoric Acid Fuel Cell (PAFC)

Phosphoric acid fuel cells (PAFC) are the oldest type with an origin that extends back to the creation of the fuel cell concept. The electrolyte used is phosphoric acid, and the cell operating temperature is about 200°C, which makes some cogeneration possible. The electrical efficiency of this cell is reasonable at about 40%. These types of fuel cells are considered too bulky for transportation applications, while higher efficiency designs exist for stationary applications.

4.1.2.5 Molten Carbonate Fuel Cell (MCFC)

Molten carbonate fuel cells, originally developed to operate directly from coal, operate at 600°C and require CO or CO_2 on the cathode side and hydrogen on the anode. The cells use carbonate as the electrolyte. The electrical efficiency of these fuel cells is high at about 50%, but the excess heat can be used for cogeneration for improved efficiency. The high temperatures required make these fuel cells not particularly suitable for vehicular applications, but they can be used for stationary power generation.

4.1.2.6 Solid Oxide Fuel Cell (SOFC, ITSOFC)

Solid oxide fuel cells (SOFCs) use a solid ionic conductor as the electrolyte rather than a solution or a polymer, which reduces corrosion problems. However, to achieve adequate ionic conductivity in such a ceramic, the system must operate at very high temperatures. The original designs, using yttria-stabilized zirconia as the electrolyte, required temperatures as high as 1000°C to operate, but the search for materials capable of serving as the electrolyte at lower temperatures resulted in the "intermediate temperature solid oxide fuel cell" (ITSOFC). This fuel cell has high electrical efficiency of 50 to 60%, and residual heat can also be used for cogeneration. Although not a good choice for vehicle applications, it is at present the best option for stationary power generation.

The fuel cell features described above are summarized in Table 4.1. The usable energy and relative costs of various fuels used in fuel cells are listed in Table 4.2. The selection of fuel cells as the primary energy source in EVs and HEVs depends on a number of issues, ranging from fuel cell technology to infrastructure to support the system. Based on the discussion in this section, the choice of fuel cell for the vehicular application is an alkaline or proton exchange design, while for stationary applications, it will be the SOFC. The size, cost, efficiency, and start-up transient times of fuel cells are yet to be at an acceptable stage for EV and HEV applications. The complexity of the controller required for fuel cell operation is another aspect that needs further attention. Although its viability has been well-proven in the space program, as well as in prototype vehicles, its immature status makes it a longer-term enabling technology for an EV and HEV.

4.1.3 HYDROGEN STORAGE SYSTEMS

The options for storage of hydrogen play a critical role in the future development of infrastructure for fuel-cell-powered EVs and hybrid vehicles. The hydrogen gas at atmospheric pressure has a fairly low energy density and is not a suitable fuel for storage. Hydrogen could be stored as compressed or liquefied gas, or in a more advanced manner by using metal hydrides or carbon nanotubes. Gas storage in compressed form is an option that has been in use for a long time. In this method,

TABLE 4.1
Fuel Cell Types

Fuel Cell Variety	Fuel	Electrolyte	Operating Temperature	Efficiency	Applications
Phosphoric acid	H_2, reformate (LNG, methanol)	Phosphoric acid	~200°C	40–50%	Stationary (>250 kW)
Alkaline	H_2	Potassium hydroxide solution	~80°C	40–50%	Mobile
Proton exchange membrane	H_2, reformate (LNG, methanol)	Polymer ion exchange film	~80°C	40–50%	EV and HEV, industrial up to ~80 kW
Direct methanol	Methanol, ethanol	Solid polymer	90–100°C	~30%	EV and HEVs, small portable devices (1 W to 70 kW)
Molten carbonate	H_2, CO (coal gas, LNG, methanol)	Carbonate	600–700°C	50–60%	Stationary (>250 kW)
Solid oxide	H_2, CO (coal gas, LNG, methanol)	Yttria-stabilized zirconia	~1000°C	50–65%	Stationary

TABLE 4.2
Usable Energy and Cost of Fuels

Fuel	Usable Energy, MJ/kg	Relative Cost/MJ
Hydrogen:		
95% pure at plant	118.3	1.0
99% pure in cylinders	120	7.4
LPG (propane)	47.4	0.5
Gasoline	45.1	0.8
Methanol	21.8	3.3
Ammonia	20.9	3.6

a large amount of energy is required to compress the gas to a level that will make storage viable, usually at a pressure of several hundred atmospheres.[4] Generation of liquid H_2 requires further compression, along with refrigeration to cryogenic temperatures, and is not likely to become a viable means of storage for vehicle applications.

Advanced methods for H_2 storage include the use of metal hydrides or carbon nanotubes. Here, the gas is compressed to a lower pressure level (a few to a few tens of atmospheres) and fed into a container filled with a material that can absorb and release H_2 as a function of the pressure, temperature, and amount of stored hydrogen in the system. The use of metal hydrides reduces the volumetric and pressure requirements for storage, because when fully loaded, these metal hydrides can contain twice as many hydrogen atoms than an equivalent volume of liquid hydrogen. The problem is that it is much heavier than the other solutions. However, current efforts are under way by several automakers to include this in the structure of the vehicle, which may result in an overall acceptable vehicle weight. The prospect of using carbon-nanotube-based materials for hydrogen storage is exciting, because it could eliminate most of the weight penalty. However, it should be noted that the properties of carbon nanotubes regarding their usefulness as H_2 storage materials is still controversial.

One of the myths that must be overcome to popularize fuel cell EVs is the safety of carrying pressurized hydrogen on board. The safety of hydrogen handling has been explored by commercial entities as well as public institutions, such as Air Products and Chemicals, Inc.[5] and Sandia National Laboratories.[6] The recommendations for its safe handling have been issued.[5] In addition, the Ford report suggested that with proper engineering, the safety of a hydrogen vehicle could be better than that of a propane or gasoline vehicle.[4]

4.1.4 REFORMERS

Many in the automotive industry have been exploring the use of methanol, ethanol, or gasoline as a fuel and reforming it on-board into hydrogen for the fuel cell. The reformer is the fuel processor that breaks down a hydrocarbon, such as methanol, into hydrogen and other by-products. The advantage of this approach is the ease of handling of hydrocarbon fuel compared to hydrogen gas, substantiated by the difficulty in storage and generation of pure hydrogen.

The accepted methods of reforming technique for vehicular fuel cells are steam reforming, partial oxidation, and autothermal processing. The two types of steam reformers in use today use methanol and natural gas as the fuel. Gasoline can also be used as the fuel, but reforming it is an expensive and complex process. Methanol is the most promising fuel for reformers, because it reforms fairly easily into hydrogen and is liquid at room temperature. A brief description of how a methanol steam reformer works is given in the following.

The steam reformer is in fact part of a fuel processor that also includes a water–gas shift reactor.[7] Steam reforming is carried out by supplying heat to a mixture of steam and fuel over a catalyst. In the case of natural gas, the mixture is fed to the reformer containing a catalyst (such as nickel) at a steam-to-carbon molar ratio of 3.5:1 and 760°C. The reformate is then passed through a low-temperature water–gas shift reactor containing a CuO-ZnO catalyst at 230°C. The process reduces the CO level to below 0.5%, which is essential since CO will poison the fuel cell anode platinum catalyst. The overall reactions in the steam reformer and the water–gas shift reactor are:

$$CH_4 + H_2O \rightarrow CO + 3H_2$$

$$CO + H_2O \rightarrow CO_2 + H_2$$

$$CH_4 + 2H_2O \rightarrow CO_2 + 4H_2$$

In the case of methanol, the mixture of methanol and fuel is reformed in the steam reformer using a catalyst of the transition metal oxide type at a steam-to-carbon molar ratio of 1.3:1 or higher and 250°C. The overall process reaction including the water–gas shift reaction is:

$$CH_3OH + H_2O \rightarrow 3H_2 + CO_2$$

The major pollutant released as exhaust from the steam reformer is the greenhouse gas carbon dioxide (CO_2), although the concentration is minimal compared to that in the exhaust of internal combustion engines.

The argument for using reformers is that the infrastructure for the production and distribution of such fuel is already in place, although widespread conversion to methanol systems is not straightforward for methanol fuels due to high corrosivity.[3] While hydrogen gas would lead to true zero-emission vehicles, it should be noted that reforming hydrocarbon fuels, including methanol and other possible biomass fuels, only shifts the source of emissions to the reformer plant. Other factors to consider are safety of methanol vs. hydrogen handling, including the fact that methanol is violently toxic, whereas hydrogen is innocuous. Methanol vapors tend to accumulate in enclosed spaces like those of gasoline leading to the formation of potentially explosive mixtures, whereas hydrogen will easily escape, even from poorly ventilated areas. The overall efficiency from the well to the wheel of methanol-based transportation will be comparable or even lower than that which can be achieved today from gasoline-based ICEVs.

4.1.5 FUEL CELL EV

A fuel cell EV consists of a fuel storage system that is likely to include a fuel processor to reform raw fuel to hydrogen, a fuel cell stack and its control unit, a power-processing unit and its controller, and the propulsion unit consisting of the electric machine and drivetrain. The fuel cell has current source type characteristics, and the output voltage of a cell is low. Several fuel cells have to be stacked in series to obtain a higher voltage level, and then the output voltage needs to be boosted in order to interface with the DC/AC inverter driving an AC propulsion motor, assuming that an AC motor is used for higher power density. The block diagram of a fuel cell EV system is shown in Figure 4.3. The voltage and current values shown in the figure are arbitrary and are included to give an idea about the typical voltage ratings at different stages of the system. The power electronic interface circuit between the fuel cell and electric motor includes the DC/DC converter for voltage boost, DC/AC inverter to supply an AC motor, microprocessor/digital signal processor for controls, and battery/capacitors for energy storage. The time constant of the fuel cell stack is much slower than that of the electrical load dynamics. A battery storage system is necessary to supply the power during transient and overload conditions and also to absorb the reverse flow of energy due to regenerative braking. The battery pack voltage rating must be high in order to interface directly with the high-voltage DC link, which means that a large number of series batteries will be needed. Alternatively, a bidirectional DC/DC converter link can interface a lower voltage battery pack and the high-voltage DC bus. The battery pack can be replaced by ultracapacitors in a

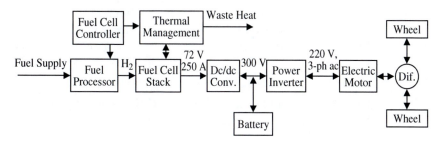

FIGURE 4.3 Fuel-cell-based EV.

fuel cell EV, although the technology is not yet ready to replace batteries. Ultra-capacitors will be discussed in the next section.

Fuel cell performance is sensitive to load variations because of the low voltage and high current output characteristics. The fuel cell controller using voltage and current feedback information regulates the flow of hydrogen into the fuel cell stack to achieve a reaction rate that delivers the required electrical power with minimum excess hydrogen vented. Attempts to draw more power out of the fuel cell without changing the flow rate deplete the concentration of hydrogen, which reduces the output voltage and may lead to damage to the fuel cell membrane.[8] The fuel cell characteristic curves as a function of flow rate are shown in Figure 4.4. When the hydrogen utilization rate approaches 100%, the cell goes into the current limit mode when it is dominated by high internal losses. The fuel cell controller must avoid operation in the current limit regime in order to maintain a decent efficiency of operation. The output power deliverability of the fuel cell stack reduces with a reduced flow rate of hydrogen, but if lower power is required for traction, then operating the fuel cell at a reduced flow rate minimizes wasted fuel. The ideal controller delivers fuel to the cell at exactly the same rate at which it is consumed by the cell to generate the electricity for the desired propulsion power. However, due to the slow response characteristics of the fuel cell, a reserve of energy is required to provide uninterrupted operation.

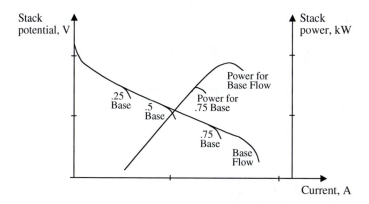

FIGURE 4.4 Fuel cell characteristics as a function of flow rate.

The by-product of the fuel cell reaction is water in the form of steam that exits the cell along with any excess hydrogen. The water vapor can be used for heating the inside of the vehicle, but the hydrogen that is vented out is a waste for the system.

EXAMPLE 4.1

The current drawn by an electric motor of a fuel cell EV for a SAE Schedule D J227A driving cycle is

$$I = \begin{cases} 9.36t + 1.482e - 3t^3 \ A & \text{for} & 0 < t < 28 \\ 61.42 \ A & \text{for} & 28 < t < 78 \\ 0 \ A & \text{otherwise} \end{cases}$$

The fuel ow rate for PEM fuel cell used in the vehicle is

$$N_f = \frac{405\,I}{nF} \ \text{gms/s}$$

(a) Calculate the amount of fuel (hydrogen) needed for one cycle of Schedule D.
(b) Calculate the range of the vehicle using Schedule D for the storage capacity of 5 kg of hydrogen.

Solution

(a) The fuel ow rate from the given equation is

$$N_f = \begin{cases} 0.01964t + 3.11e - 6t^3 & 0 < t < 28 \\ 0.1289 & 28 < t < 78 \\ 0 & \text{otherwise} \end{cases}$$

The amount of fuel (hydrogen) needed in one cycle can be obtained by integrating the $N_f(t)$ for $t = 0$ s to $t = 78$ s.

Using numerical integration, we can show that

$$\int_0^{78} N_f(t)\,dt = 15.10 \ \text{gms}$$

(b) For Schedule D, one cycle is equivalent to 1 mi. Therefore, 5 kg of hydrogen will give a range of 5000/15.1 = 331 mi.

4.2 SUPERCAPACITORS AND ULTRACAPACITORS

Capacitors are devices that store energy by the separation of equal positive and negative electrostatic charges. The basic structure of a capacitor consists of two conductors, known as plates, separated by a dielectric, which is an insulator. The power densities of conventional capacitors are extremely high ($\sim 10^{12}$ W/m^3), but the energy density is very low (~ 50 Wh/m^3).[9] These conventional capacitors are commonly known as "electrolytic capacitors." They are widely used in electrical circuits as intermediate energy storage elements for time constants that are of a completely different domain and are of much smaller order compared to the energy storage devices that are to serve as the primary energy sources for EVs. The capacitors are described in terms of capacitance, which is directly proportional to the dielectric constant of the insulating material and inversely proportional to the space between the two conducting plates. The capacitance is measured by the ratio of the magnitude of the charge between either plate and the potential difference between them ($C = q/V$).

Supercapacitors and ultracapacitors are derivatives of conventional capacitors, where energy density has been increased at the expense of power density to make the devices function more like a battery. Power density and energy density of supercapacitors and ultracapacitors are of the order of 10^6 W/m^3 and 10^4 Wh/m^3, respectively. Energy density is much lower compared to those of batteries (~ 5 to 25×10^4 Wh/m^3), but the discharge times are much faster (110 s compared to $\sim 5 \times 10^3$ s of batteries), and the cycle life is much more ($\sim 10^5$ compared to 100 to 1000 of batteries).[9–11]

Supercapacitors contain an electrolyte that enables the storage of electrostatic charge in the form of ions, in addition to conventional energy storage in electrostatic charges, like in an electrolytic capacitor. The internal functions in a supercapacitor do not involve electrochemical reaction. The electrodes in supercapacitors are made of porous carbon with high internal surface area to help absorb the ions and provide a much higher charge density than is possible in a conventional capacitor. The ions move much more slowly than electrons, enabling a much longer time constant for charging and discharging compared to electrolytic capacitors.

Ultracapacitors are versions of electrolytic capacitors that use electrochemical systems to store energy in a polarized liquid layer at the interface between an ionically conducting electrolyte and an electrically conducting electrode. Energy storage capacity is increased by increasing the surface area of the interface, similar to that in a supercapacitor. Electrochemical (also known as Faradaic) reactions in ultracapacitors are confined to the surface layers and, hence, are fully reversible with a long cycle life.

Current research and development aim to create ultracapacitors with capabilities in the vicinity of 4000 W/kg and 15 Wh/kg. The possibility of using supercapacitors and ultracapacitors as primary energy sources is quite far reaching, although it is likely that these can be improved to provide sufficient energy storage in HEVs. On the other hand, supercapacitors and ultracapacitors with high specific power are suitable as an intermediate energy transfer device in conjunction with batteries or fuel cells in EVs and HEVs to provide sudden transient power demand, such as during acceleration and hill climbing. The devices can also be used efficiently to capture recovered energy during regenerative braking.

4.3 FLYWHEELS

The flywheel is the kind of energy supply unit that stores energy in mechanical form. Flywheels store kinetic energy within a rotating wheel-like rotor or disk made of composite materials. Flywheels have a long history of usage in automobiles, being routinely used in all of today's IC engines to store energy and smooth the power delivered by abrupt pulses of the engine. However, the amount of energy storage required in flywheels of IC engines is small and is limited by the need of the vehicle to accelerate rapidly. The flywheel is currently being looked into for use in a number of different capacities. Flywheels can be used in HEVs with a standard IC engine as a power assist device. Alternatively, flywheels can be used to replace chemical batteries in EVs to serve as the primary energy source or could be used in conjunction with batteries. However, technological breakthroughs in increasing the specific energy of flywheels are necessary before they can be considered as the energy source for EVs and HEVs. The flywheels of today are quite complex, large, and heavy. Safety is also a concern with flywheels.

The flywheel design objective is to maximize energy density. The energy U stored in the flywheel is given by

$$U = \frac{1}{2} J \omega^2$$

where J is the polar moment of inertia, and ω is the angular velocity. Energy storage is increased by spinning at higher velocities without increasing the inertia, which is directly proportional to mass. Increasing angular velocity ω, in turn, increases centrifugal stress, which must not exceed failure stress with a given factor of safety. Stored energy per unit mass can be expressed as follows:

$$\frac{U}{m} = k \frac{\sigma}{\rho}$$

where k is a constant depending on the geometry, σ is the tensile strength, and ρ is the density of the material. Therefore, the material to be used in a flywheel must be lightweight with high tensile strength, conditions that are satisfied by composite materials.

Flywheels have several advantages as an energy source, the most important of which is the high specific power. Theoretically, specific power of flywheels has been shown to be of the order of 5 to 10 kW/kg, with a specific power of 2 kW/kg being easily achievable without exceeding safe working stresses. Other performance features that make flywheels attractive can be attributed to their mechanical nature. Flywheels are not affected by temperature extremes. There are no concerns with toxic chemical processing and disposal of waste materials, making flywheels environmentally friendlier than chemical batteries. Flywheel energy storage is reliable in that it possesses excellent controllability and repeatability characteristics. The state of charge in flywheels is precisely known at all times through measurement of the rotational speed. The energy conversion process to and from the flywheel approaches 98%, compared to 75 to 80% of batteries. The service life of a flywheel is many times that of a battery, with little maintenance required. The charging of flywheels is a fraction of that required by batteries and can be less than 10 min for full recharge in a flywheel charging station. The ability to absorb or release a high amount of power in a short period of time also aids the regenerative braking process.

Despite several advantages, there are still a number of significant drawbacks with flywheels. The major difficulty in implementing a flywheel energy storage system is in the extra equipment needed to operate and contain the device. The extras are particularly difficult in EV and HEV applications, where the extra weight and expense make a big difference. In order to reduce windage losses, the flywheel needs to be enclosed in a vacuum chamber. The vacuum condition adds additional constraints on the bearings, because liquid-lubricated bearings do not survive in vacuum. The alternative is to use magnetic bearings, which are in a development stage. The biggest extra weight in flywheels comes from the safety containment vessel, which is required for protection from the dangerous release of sudden energy and material in the case of a burst failure.

A flywheel, similar to a battery, goes through charge and discharge processes in order to store and extract energy, which earned it the name "electromechanical battery." The rotor's shaft is coupled with a motor and generator, which, during charging, spin the rotor to store the kinetic energy and, during discharging, convert the stored energy into electric energy. Interface electronics is necessary to condition the power input and output and to monitor and control the flywheel. Modern flywheels are made of composite materials, such as carbon fiber, instead of steel to increase the energy density, which can be up to 200 Wh/kg. A composite material flywheel has the additional advantage in that it disintegrates in the form of a fluid, as compared to large metallic pieces for a steel-made flywheel, in the case of a catastrophic burst.

REFERENCES

1. Appleby, A.J. and Foulkes, F.R., *Fuel Cell Handbook*, Van Nostrand Reinhold, New York, 1989.
2. Andrews, N., Poised for growth: DG and ride through power, *Power Quality*, January/February, 10–15, 2002.
3. Laughton, M.A., Fuel cells, *Power Eng. J.*, February, 37–47, 2002.
4. Ford Motor Co., Direct-Hydrogen-Fueled Proton-Exchange-Membrane Fuel Cell System for Transportation Applications: Hydrogen Vehicle Safety, Report DOE/CE/50389-502, Directed Technologies Inc., Arlington, VA, May, 1997.
5. Linney, R.E. and Hansel, J.G., Safety considerations in the design of hydrogen-powered vehicles, Part 2, *Hydrogen Energy Progress XI: Proc. 11th World Hydrogen Energy Conf.*, Stuttgart, Veziroglu, T.N. et al., Eds., International Association for Hydrogen Energy, Coral Gables, FL, 1996.

6. Ringland, J.T. et al., Safety Issues for Hydrogen Powered Vehicles, Report SAND-94-8226, UC407, Sandia National Laboratories, Albuquerque, NM, March, 1994.
7. Singh, R., Will developing countries spur fuel cell surge?, *Chem. Eng. Process,* March, 59–66, 1999.
8. EC&G Services, Parson's Inc., *Fuel Cell Handbook*, 5th ed., U.S. Department of Energy, Office of Fossil Energy, October, 2000.
9. Dell, R.M. and Rand, D.A.J., *Understanding Batteries*, Royal Society of Chemistry, Cambridge, UK, 2001.
10. Dhameja, S., *Electric Vehicle Battery Systems*, Newnes (Elsevier Science), Burlington, MA, 2002.
11. Rand, D.A.J., Woods, R., and Dell, R.M., *Batteries for Electric Vehicles*, John Wiley & Sons, New York, 1998.

PROBLEM

4.1

The current drawn by an electric motor of a fuel cell EV for a SAE Schedule D J227A driving cycle is

$$I = \begin{cases} 9.5t + 1.5A & \text{for} & 0 < t < 28 \\ 55A & \text{for} & 28 < t < 78 \\ 0A & \text{otherwise} \end{cases}$$

The fuel flow rate for PEM fuel cell used in the vehicle is

$$N_f = \frac{405\,I}{nF} \text{ gms/sec}$$

(a) Calculate the amount of fuel (hydrogen) needed for one cycle of Schedule D.

(b) Calculate the amount of hydrogen needed for a range of 200 mi.

5 DC and AC Electric Machines

The electric machine delivers processed power or torque to the transaxle to propel the vehicle. The machine also processes the power flow in the reverse direction during regeneration, when the vehicle is braking, converting mechanical energy from the wheels into electrical energy. The term "motor" is used for the electric machine when energy is converted from electrical to mechanical, and the term "generator" is used when power flow is in the opposite direction, with the machine converting mechanical energy into electrical energy. The braking mode in electric machines is referred to as regenerative braking.

In electric vehicles (EVs), the electric motor is the sole propulsion unit, while in hybrid electric vehicles (HEVs), the electric motor and the internal combustion (IC) engine together in a series or parallel combination provide the propulsion power. In an EV or an HEV, the electric traction motor converts electrical energy from the energy storage unit to mechanical energy that drives the wheels of the vehicle. The major advantages of an electric motor over an IC engine are that the motor provides full torque at low speeds and the instantaneous power rating can be two or three times the rated power of the motor. These characteristics give the vehicle excellent acceleration with a nominally rated motor.

Electric motors can be DC type or AC type. The DC series motors were used in a number of prototype EVs in the 1980s and prior to that due to their developed status and ease of control. However, the size and maintenance requirements of DC motors are making their use obsolete, not just in the automotive industry, but in all motor drive applications. The more recent EVs and HEVs employ AC and brushless motors, which include induction motors, permanent magnet motors, and switched reluctance motors. The AC induction motor technology is quite mature, and significant research and development activities have taken place on induction motor drives over the past 50 years. The control of induction motors is more complex than DC motors, but with the availability of fast digital processors, computational complexity can be easily managed. Vector control techniques developed for AC motor drives make the control of AC motors similar to that of DC motors through reference frame transformation techniques. The computational complexity arises from these reference frame transformations, but today's digital processors are capable of completing complex algorithms in a relatively short time.

The competitor to the induction motor is the permanent magnet (PM) motor. The permanent magnet AC motors have magnets on the rotor, while the stator construction is the same as that of the induction motor. The PM motors can be of surface-mounted type, or the magnets can be inset within the rotor in the interior

PM motors. The PM motor can also be classified as sinusoidal type or trapezoidal type depending on the flux distribution in the air gap. Trapezoidal motors have concentrated three-phase windings and are also known as brushless DC motors. The PM motors are driven by a six-switch inverter just like an induction motor, but the control is relatively simpler than that of the induction motor. The use of high-density rare earth magnets in PM motors provides high power density, but at the same time, the cost of magnets is on the negative side for these motors. For EV and HEV applications, motor size is relatively large compared to the other smaller power applications of PM motors, which amplifies the cost problem. However, HEV motors are usually smaller than EV motors, and the performance and efficiency achievable from PM motors may be enough to overcome the cost problem. Interior PM motors have excellent performance characteristics, superior to surface mount magnets, but manufacturing complexity is one of the drawbacks for these motors.

Another candidate for traction motors is the switched reluctance (SR) motor. These motors have excellent fault tolerance characteristics, and their construction is fairly simple. The SR motors have no windings, magnets, or cages on the rotor, which helps increase the torque and inertia rating of these motors. The motor speed–torque characteristics are an excellent match with the road load characteristics, and performance of SR motors for EV/HEV applications have been found to be excellent. The two problems associated with SR motors are acoustic noise and torque ripple. The noise problems of SR motors are being addressed in various research activities.

5.1 MOTOR AND ENGINE RATINGS

The strength of electric motors and IC engines are typically described with "hp" or horsepower ratings, although a comparison between electric motors and IC engines in terms of hp only is not fair. The power that an electric motor can continuously deliver without overheating is its rated hp, which is typically a derated figure. For short periods of time, the motor can deliver two to three times the rated hp. Therefore, at starting, high power is available from an electric motor for acceleration, and the motor torque can be maximum under stall conditions, i.e., at zero speed. Motor type determines whether maximum torque is available at zero speed or not. On the contrary, an IC engine is rated at a specific r/min level for maximum torque and maximum hp. The IC engine maximum torque and hp ratings are typically derived under idealized laboratory conditions. In practical situations, it is impossible to achieve the rated hp; the maximum hp available from an IC engine is always smaller than the rated hp.

The torque characteristics of motors are shown in Figure 5.1, along with torque characteristics of IC engines. The characteristics of specific motors and IC engines will differ somewhat from these generalized curves. For electric motors, a high torque is available at starting, which is the peak torque of the motor. The peak torque is much higher (typically twice) than that of the rated torque. The peak torque for electric motors in an EV/HEV application needs to be sustained for about 60 to 90 s. The *rated torque* for electric motors is the torque that must be available for an indefinite length of time. The peak or rated power from a motor is obtained at base speed (ω_b) when motor characteristics enter the constant power region from the constant

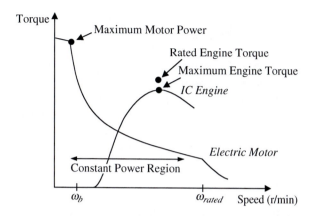

FIGURE 5.1 Electric motor and IC engine torque characteristics.

torque region, once the voltage limit of the power supply is reached. The motor rated speed (ω_{rated}) is at the end of the constant power region. The IC engine peak power and torque occur at the same speed. At this stage, it is helpful to review the power and torque relation, which is as follows:

$$\text{Power}\,(\text{Watts}) = \text{Torque}\,(\text{N}-\text{m}) \times \text{speed}\,(\text{rads/s}) \qquad (5.1)$$

The power–torque relation in hp and ft-lb is

$$HP = \frac{\text{Torque}\,(\text{ft}-\text{lbs}) \times \text{rpm}}{5232}$$

Figure 5.1 depicts that the IC engine does not produce torque below a certain speed. A transmission is essential for an IC engine to match the vehicle speed with the narrow high hp speed range of the engine. On the other hand, the electric motor produces high torque even at zero speed and typically has constant power characteristics over a wide speed range. Therefore, the electric motor can be attached directly to the drive wheels and accelerate the vehicle from zero speed all the way to top speed. The motor and, hence, the vehicle speed can be controlled directly through the power electronic converter feeding the current into the motor. There is essentially no requirement for a transmission with an electric motor, other than a fixed gear for appropriately sizing the motor.

5.2 EV AND HEV MOTOR REQUIREMENTS

The important characteristics of a motor for an EV or HEV include flexible drive control, fault tolerance, high efficiency, and low acoustic noise. The motor drive must be capable of handling voltage fluctuations from the source. Another important requirement of the electric motor is acceptable mass production costs, which is to be achieved through technological advancement. The requirements of an EV or HEV motor, not necessarily in order of importance, are itemized in the following:

- Ruggedness
- High torque-to-inertia ratio (T_e/J); large T_e/J results in "good" acceleration capabilities
- Peak torque capability of about 200 to 300% of continuous torque rating
- High power-to-weight ratio (P_e/w)
- High-speed operation, ease of control
- Low acoustic noise, low electromagnetic interference (EMI), low maintenance, and low cost
- Extended constant power region of operation

5.3 DC MACHINES

The torque in electric machines is produced utilizing one of two basic principles of electromagnetic theory: by Lorentz force principle, where torque is produced by the mutual interaction of two orthogonal magnetomotive forces (mmf); and by reluctance principle, where the rotor produces torque while moving toward the minimum reluctance position in a varying reluctance path. The DC and AC machines, including the permanent magnet machines, work on the first principle, while the switched reluctance machines work on the latter principle.

The DC machines have two sets of windings, one in the rotor and the other in the stator, which establish the two fluxes; hence, the mmfs that interact with each other produce the torque. The orthogonality of the two mmfs, which is essential for maximum torque production, is maintained by a set of mechanical components called commutators and brushes. The winding in the rotor is called the armature winding, while the winding in the stationary part of the machine is called the field winding. The armature and the field windings are supplied with DC currents. The armature windings carry the bulk of the current, while the field windings carry a small field excitation current. The armature and the field currents in the respective windings establish the armature and field mmfs. The magnitude of the mmfs is the product of the number of turns in the windings and the current. Depending on the number of supply sources and the type of connection between the armature and field windings, there can be several types of DC machines. When the armature and field windings are supplied from independently controlled DC sources, then it is known as a separately excited DC machine. The separately excited DC machine offers the maximum flexibility of torque and speed control through independent control of the armature and field currents. The DC shunt machine has a similar parallel configuration of armature and field windings as in the separately excited motor, except that the same DC source supplies both armature and field windings. In shunt motors, simplicity in the power supply is compromised for reduced flexibility in control. In another type of DC machine, known as the series DC machine, the armature and the series windings are connected in series, and the machine is supplied from a single source. Because the armature and the field windings carry the same current, the field is wound with a few turns of heavy gauge wire to deliver the same mmf or ampere-turns as in the separately excited machine. The greatest advantage of the series machine is the high starting torque that helps achieve rapid acceleration. However, control flexibility is lost due to the series connection of armature and field windings.

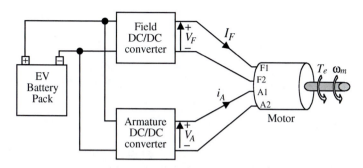

FIGURE 5.2 A DC motor drive, including power electronics and battery source.

Positive attributes of DC machines are as follows:

- Ease of control due to linearity
- Capability for independent torque and flux control
- Established manufacturing technology

Disadvantages of DC machines include the following:

- Brush wear that leads to high maintenance
- Low maximum speed
- EMI due to commutator action
- Low power-to-weight ratio
- Low efficiency

The separately excited DC motor used in an EV or HEV has two separate DC/DC converters supplying the armature and field windings from the same energy source, as shown in Figure 5.2. The DC/DC converters process the fixed supply voltage of the energy source to deliver a variable DC to the armature and field circuits. The power rating of the converter supplying the armature windings is much larger than that of the converter supplying the field winding. Control inputs to the converter circuits are the desired torque and speed of the motor. Control outputs of the converters are the voltages applied to the armature and field circuits of the DC motor.

Independent armature voltage and field current control, possible in separately excited DC machines or motors, offer the possibility of additional performance optimization in addition to meeting the torque–speed requirements of the machine. The indices used for measuring performance in motor drives include efficiency, torque per ampere, torque ripple, response time, etc. The weights on the individual performance indices depend on the application and the design requirements. The most critical performance index for EV and hybrid vehicle applications is efficiency. The analysis to follow on DC motors, based on separately excited DC motors, is intended to set forth the premise for performance analysis of DC drives in the next chapter.

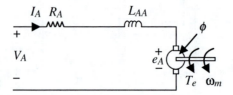

FIGURE 5.3 Armature equivalent circuit of a DC motor.

The armature equivalent circuit of a DC motor is shown in Figure 5.3. The circuit consists of the armature winding resistance R_A, the self-inductance of armature winding L_{AA} and the back-emf e_A. The variables shown in the figure are as follows:

V_A = armature voltage
I_A = armature current
T_e = developed motor torque
ω_m = shaft speed
ϕ = armature linking flux (primarily from field current)

Applying *KVL* around the armature circuit, the voltage balance equation is

$$V_A = R_A i_A + L_{AA} \frac{di_A}{dt} + e_A \tag{5.2}$$

where

$$e_A = K\phi\omega_m$$
$$T_e = K\phi i_A \tag{5.3}$$

e_A is known as the back-emf, and K is a machine constant that depends on the machine construction, number of windings, and core material properties. The field equivalent circuit of the DC motor is shown in Figure 5.4. The field circuit consists of the field winding resistance R_F and the self-inductance of the field winding L_{FF}. The voltage applied to the field is V_F. The field circuit equation is

$$V_F = R_F i_F + L_{FF} \frac{di_F}{dt}$$

The resistances of the field windings in separately excited and shunt DC motors are high, because there are a lot of turns in the winding. The transient response in the field circuit is, thus, much faster than the armature circuit. The field voltage is also typically not adjusted frequently, and for all practical purposes, a simple resistor fed from a DC source characterizes the electrical unit of the field circuit. The field current establishes the mutual flux or field flux, which is responsible for torque

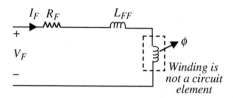

FIGURE 5.4 DC motor field equivalent circuit.

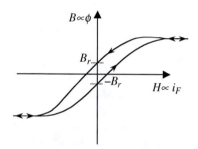

FIGURE 5.5 Typical DC motor magnetization characteristic.

production in the machine. The field flux is a nonlinear function of field current and can be described by

$$\phi = f\left(i_F\right)$$

The electromagnetic properties of the machine core materials are defined by the following relationship:

$$B = \mu H$$

where B is the magnetic flux density in Tesla or weber/m^2, H is the magnetic field intensity in ampere-turn/m, and μ is the permeability of the material. The permeability, in turn, is given by $\mu = \mu_0 \mu_r$, where $\mu_0 = 4\pi \times 10^{-7}$ H/m is the permeability of free space, and μ_r is the relative permeability. The relative permeability of air is one. The B–H relationship of magnetic materials is nonlinear and is difficult to describe by a mathematical function. Likewise, the field circuit of DC machines is characterized by nonlinear electromagnetic properties of the core, which is made of ferromagnetic materials. The properties of core materials are often described graphically in terms of the B–H characteristics, as shown in Figure 5.5. Nonlinearity in the characteristics is due to the saturation of flux for higher currents and hysteresis effects. When an external magnetization force is applied through the currents in the windings, the magnetic dipole moments tend to align to orient in a certain direction. This dipole orientation establishes a large magnetic flux, which would not exist without the external magnetization force applied on the core. The magnetic dipole

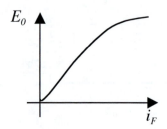

FIGURE 5.6 Magnetization characteristics of DC machines.

moments relax toward their random orientation upon removal of the applied magnetic force, but few dipole moments retain their orientation in the direction of the previously existing magnetization force. The retention of direction phenomenon of the dipole moments is known as magnetic hysteresis. Hysteresis causes magnetic flux density B to be a multivalued function that depends on the direction of magnetization. The magnetic effect that remains in the core after the complete removal of magnetization force is known as the residual magnetism (denoted by B_r in Figure 5.5). The direction of the residual flux, as mentioned previously, depends on the direction of field current change. The $B–H$ characteristics can also be interpreted as the $\phi–i_F$ characteristics, because B is proportional to ϕ, and H is proportional to i_F for a given motor. Saturation in the characteristics reflects the fact that no more magnetic dipole moments remain to be oriented once sufficient magnetization force has been applied and the flux has reached the maximum or saturation level.

The energy required to cause change in magnetic orientations is wasted in the core material and is referred to as hysteresis loss. The area of the hysteresis loop in magnetization characteristics is proportional to hysteresis loss.

For most applications, it is sufficient to show the magnetic properties of core materials through a single-valued, yet nonlinear, function, which is known as the DC magnetization curve. The magnetization curve of a DC machine is typically shown as a curve of open-circuit-induced voltage E_0 vs. field current i_F at a particular speed. The open-circuit-induced voltage is nothing but the back-emf e_A, which is linearly proportional to the flux at a constant speed (refer to Figure 5.3 and Equation 5.3). Therefore, the shape of this characteristic, shown in Figure 5.6, is similar to that of the magnetic characteristics of the core material.

The torque–speed relationship of a DC motor can be derived from Equations 5.2 and 5.3 and is given by

$$\omega_m = \frac{V_A}{K\phi} - \frac{R_A}{(K\phi)^2} T_e \tag{5.4}$$

The torque–speed characteristic is shown in Figure 5.7. The positive torque axis represents the motoring characteristic, while the negative torque region represents the generating characteristic. The speed–torque characteristic is adjusted through the armature voltage or the field current. For a given speed and torque (i.e., a point (T^*, ω_m^*) in the $\omega – T$ plane), there are an infinite number of corresponding armature

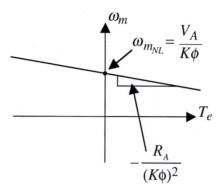

FIGURE 5.7 Speed–torque characteristics of a DC motor.

voltages and field currents, as shown in Figure 5.8a, that would satisfy Equation 5.4. A smart control design will optimize one or more performance indices and operate the motor on the optimized characteristic curve. To follow up on the concept, let us assume that the controller can set the field current and the armature voltage, and we are interested in minimizing the loss in the machine. The driver input commands set the desired torque and speed (T^*, $\omega_m{}^*$) of the machine. Inserting the operating point in the armature voltage equation, we have

$$V_A = R_A I_A^* + K\phi\omega_m{}^*$$

$$\Rightarrow V_A = \frac{R_A}{K\phi}T^* + K\phi\omega_m{}^* \quad \text{or} \quad (5.5)$$

$$\omega_m{}^* K^2\phi^2 - V_A K\phi + R_A T^* = 0$$

Equation 5.5 gives all possible combinations of armature voltage and field flux that will give the same operating point (T^*, $\omega_m{}^*$). The possible combinations are shown graphically in Figure 5.8b. The optimization algorithm will select the right combination of V_A and ϕ that will minimize losses. The loss in DC machines is minimized when armature circuit dependent losses equal the field circuit dependent losses.[1] Knowing the machine parameters, V_A and ϕ commands can be set such that the armature circuit losses equal the field circuit losses to minimize the overall loss, hence, maximizing efficiency.

5.4 THREE-PHASE AC MACHINES

The primary difference between AC machines and DC machines is that the armature circuit of the former is located in the stationary piece of the structure. The major advantage of this arrangement is the elimination of the commutator and brushes of DC machines. The machines are fed from AC sources and can be single-phase or multiple-phase types. Single-phase AC machines are used for low-power appliance applications, while higher-power machines are always of three-phase configuration.

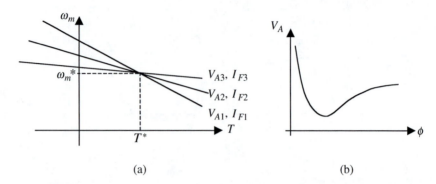

(a)   (b)

FIGURE 5.8 (a) Steady state operating point. (b) Required armature voltage vs. flux at a fixed operating point.

The second mmf required for torque production in AC machines (equivalent to the field mmf of DC machines) comes from the rotor circuit. Depending on the way the second mmf is established, AC machines can be of induction type or synchronous type. For either of the two types of AC machines, the stator windings are identical in nature.

5.4.1 SINUSOIDAL STATOR WINDINGS

The three-phase stator windings of AC machines are sinusoidally distributed spatially along the stator circumference, as shown in Figure 5.9a, to establish a sinusoidal mmf waveform. Although the windings are shown as concentrated in locations aa', bb', and cc' for the three phases, the number of turns for each of the phase windings should vary sinusoidally along the stator circumference. This theoretical space sinusoidal distribution of Phase-a winding is shown in Figure 5.9b, which has been represented by an equivalent concentrated winding a-a' in Figure 5.9a. In practice, this sinusoidal winding distribution is achieved in a variety of ways. The number of turns per slot (accounting for turns of all the phases) is kept the same to maintain manufacturing simplicity. The equivalent distribution of the Phase-a windings in a horizontally laid stator axis as if the stator cross-section was split along the radius at $\theta = 0$ and developed longitudinally is shown in Figure 5.10a. The current passing through these Phase-a stator windings causes a sinusoidal Phase-a mmf $F_a(\theta)$, which is shown in Figure 5.10b. The mmf primarily exists in the air gap due to high permeability of the stator and rotor steel and tend to be radial in direction due to the short length of air gap relative to stator inside diameter. The sinusoidal distribution of the windings can be expressed as

$$n_{as}(\theta) = N_P \sin\theta, \quad 0 \leq \theta \leq \pi$$
$$= -N_P \sin\theta, \quad \pi \leq \theta \leq 2\pi$$

where N_P is the maximum conductor density expressed in turns per radian. Let the a-phase winding have an N_s equivalent number of turns (i.e., $2N_s$ conductors), which would give the same fundamental sinusoidal component as the actual winding

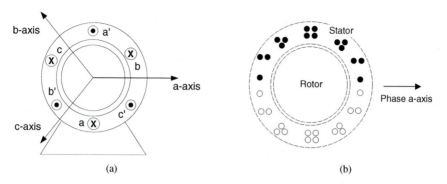

(a) (b)

FIGURE 5.9 (a) Three-phase winding and magnetic axes of an AC machine. (b) Sinusoidal distribution of Phase-a winding along the stator circumference.

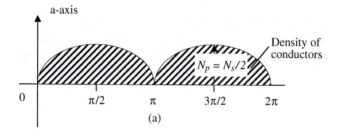

(a)

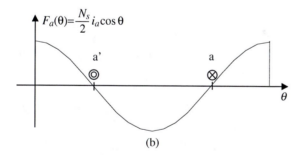

(b)

FIGURE 5.10 (a) Equivalent distribution of Phase-a winding. (b) mmf distribution of Phase a.

distribution. Therefore, the integral of the conductor density in Figure 5.10b between 0 and π has a total of N_s conductors (accounting for half the turns in a winding-half), which is

$$N_s = \int_\pi^0 N_P \sin\theta d\theta = 2N_P$$

$$\Rightarrow N_P = \frac{N_s}{2}$$

The sinusoidal conductor–density distribution in Phase-a winding is

$$n_S(\theta) = \frac{N_s}{2}\sin\theta, \quad 0 \le \theta \le \pi \tag{5.6}$$

The equivalent conductor-density is used to calculate the air gap magnetic field parameters, which consist of field intensity, flux density, and mmf. The basic relationship between magnetic field intensity H and current i is given by Ampere's Law, which states that the line integral of H around a closed path is equal to the net current enclosed

$$\left(\oint H.dl = \sum Ni \right)$$

$\sum Ni$ is the ampere-turn product defining the net current enclosed and is known as the total mmf in magnetic circuit terms. The radial magnetic field intensity H_a in the AC machine under discussion is established in the air gap, when current i_a flows through the Phase-a windings, which can be derived using Ampere's Law as

$$H_a(\theta) = \frac{N_s}{2l_g}i_a\cos\theta$$

where l_g is the effective length of the air gap. The flux density $B_a(\theta)$ and mmf $F_a(\theta)$ can be derived as

$$B_a(\theta) = \mu_a(\theta)H_a(\theta) = \frac{\mu_0 N_s}{2l_g}i_a\cos\theta$$

where μ_0 is the permeability of free space or air and

$$F_a(\theta) = l_g H_a(\theta) = \frac{N_s}{2}i_a\cos\theta \tag{5.7}$$

The mmf, flux intensity, and field intensity are all $90°$ phase shifted in space with respect to the winding distribution. The angle θ is measured in the counterclockwise direction with respect to the Phase-a magnetic axis. The field distribution shown in Figure 5.10b is for positive current. Irrespective of the direction of current, the peak of the mmf (positive or negative) will always appear along the Phase-a magnetic axis, which is the characteristic of mmf produced by a single-phase winding.

5.4.2 Number of Poles

The two equivalent Phase-a conductors in Figure 5.9a represent two poles of the machine. Electric machines are designed with multiple pairs of poles for efficient

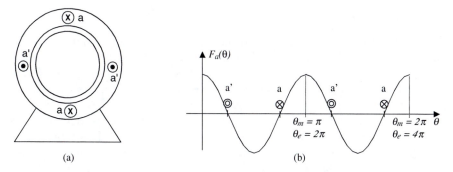

FIGURE 5.11 (a) Cross-section of a four-pole machine showing Phase-a windings only. (b) Phase-a mmf distribution.

utilization of the stator and rotor magnetic core material. In multiple pole pair machines, the electrical and magnetic variables (like induced voltages, mmf, and flux density) complete more cycles during one mechanical revolution of the motor. The electrical and mechanical angles of revolution and the corresponding speeds are related by

$$\theta_e = \frac{P}{2}\theta_m$$

$$\omega_e = \frac{P}{2}\omega_m$$

(5.8)

where P is the number of poles. The four-pole machine cross-section is represented as shown in Figure 5.11a, while the Phase-a mmf F_a as a function of θ_e or θ_m is illustrated in Figure 5.11b. The mmf is mathematically represented as

$$F_a(\theta_e) = \frac{N_s}{P}i_a\cos(\theta_e)$$

(5.9)

5.4.3 THREE-PHASE SINUSOIDAL WINDINGS

Phases b and c have sets of windings that are similar to the Phase-a winding described in the previous section, except that they are displaced 120° spatially with respect to each other, as shown in Figure 5.9a. The resulting mmfs due to currents in Phases b and c can be expressed as

$$F_b(\theta) = \frac{N_s}{2}i_b\cos\left(\theta - \frac{2\pi}{3}\right)$$

$$F_c(\theta) = \frac{N_s}{2}i_c\cos\left(\theta + \frac{2\pi}{3}\right)$$

(5.10)

5.4.4 SPACE VECTOR REPRESENTATION

The extensive amount of coupling existing among the circuits of three-phase AC machines makes analysis a formidable task. Axes transformations or reference frame theory is necessary to decouple the voltage expressions of the phases as well as to implement control algorithms that achieve the best performance. *Space vector representation* is a convenient method of expressing the equivalent resultant effect of the sinusoidally space distributed electrical and magnetic variables in AC machines, in a way that is similar to the use of phasors in describing the sinusoidally time-varying voltages and currents in electrical circuits. Space vectors provide a useful and compact form of representing machine equations, which not only simplifies the representation of three-phase variables but also facilitates the transformation between three- and two-phase variables. The two-phase system is an equivalent representation of the three-phase variables in a *dq* (two-axis) coordinate system, which is necessary for control implementation. The *dq* coordinate system will be elaborated in Section 5.7.

The concept of reference frame transformations originates from Parks transformations,[2] which provided a revolutionary new approach of analyzing three-phase electric machines by transforming three-phase variables (voltages, currents, and flux linkages) into two-phase variables, with the help of a set of two fictitious windings (known as *dq* windings) rotating with the rotor. The notations of space vector evolved later as a compact set of representation of the three-phase machine variables, either in the three-phase *abc* reference frame or in the fictitious two-phase *dq* reference frame.[3-5] Space vectors are more complex than phasors, because they represent time variation as well as space variation. Space vectors, just like any other vectors, have a magnitude and an angle, but the magnitude can be time varying. For example, the stator mmfs of the three phases in the AC machine can be represented by space vectors as

$$\vec{F}_a(t) = \frac{N_S}{2} i_a(t) \angle 0°$$

$$\vec{F}_b(t) = \frac{N_S}{2} i_b(t) \angle 120°$$

$$\vec{F}_c(t) = \frac{N_S}{2} i_c(t) \angle 240°$$

(5.11)

Note that space vectors are complex numbers, and "→" is used to denote the vector characteristic. Time dependence is also explicitly shown. The magnitude of the vector represents the positive peak of the sinusoidal spatial distribution, and the angle represents the location of the peak with respect to the Phase-a magnetic axis (chosen by convention). Space vectors of individual phases can now be added conveniently by vector addition to give the resultant stator mmf as follows:

$$\vec{F}_S(t) = \vec{F}_a(t) + \vec{F}_b(t) + \vec{F}_c(t) = \hat{F}_S \angle \theta_F$$

(5.12)

where $\hat{F}_S$ is the stator mmf space vector amplitude, and θ_F is the spatial orientation with respect to the Phase-a reference axis.

In general, if f represents a variable (mmf, flux, voltage, or current) in a three-phase AC machine, the corresponding resultant space vector can be calculated as

$$\vec{f}_{abc}(t) = f_a(t) = af_b(t) + a^2 f_c(t)$$

where $f_a(t)$, $f_b(t)$, and $f_c(t)$ are the magnitudes of the phase space vectors of the variables, and a and a^2 are spatial operators that handle the 120° spatial distribution of the three windings, one with respect to the other, along the stator circumference. The operators a and a^2 are $a = e^{j2\pi/3}$ and $a^2 = e^{j4\pi/3}$ and, hence, the space vector can also be represented as

$$\vec{f}_{abc}(t) = f_a(t) + f_b(t)\angle 120 + f_c(t)\angle 240 \tag{5.13}$$

The space vector can be used to represent any of the AC machine sinusoidal variables either in the stator circuit or in the rotor circuit. For example, the flux-density, current, and voltage space vectors of the stator can be expressed as,

$$\vec{B}_S(t) = \frac{\mu_0 N_S}{2l_g} i_a(t) + \frac{\mu_0 N_S}{2l_g} i_b(t)\angle 120 + \frac{\mu_0 N_S}{2l_g} i_c(t)\angle 240 = \hat{B}_S \angle \theta_B$$

$$\vec{i}_S(t) = i_a(t) + i_b(t)\angle 120 + i_c(t)\angle 240 = \hat{I}_S \angle \theta_I \tag{5.14}$$

$$\vec{v}_S(t) = v_a(t) + v_b(t)\angle 120 + v_c(t)\angle 240 = \hat{V}_S \angle \theta_V$$

The space vector $\vec{f}_{abc}$ for a balanced set of three-phase variables f_a, f_b, and f_c has a magnitude 3/2 times greater than the magnitude of the phase variables, and its spatial orientation is at an angle ωt at time t, with respect to the reference Phase-a axis. Here, ω is the angular frequency of the phase variables. Therefore, the amplitude of the space vector is constant for a balanced set of variables, but the phase angle (i.e., the spatial orientation) is a function of time. The space vector at any instant of time can be obtained by the vector sum of the three-phase variables, as shown in Figure 5.12 for stator currents. Note that there is a unique set of phase variables that would sum up to give the resultant space vector $\vec{f}_{abc}$, because $f_a + f_b + f_c = 0$ for a balanced set of variables. The multiplication factor of 2/3 in the figure represents the unique relationship between the magnitude of the space vector and the peak magnitude of the time varying phase sinusoids. Examples with numerical values are given in the following to supplement the theory.

EXAMPLE 5.1

The stator currents of a three-phase machine at $\omega t = 40°$ are as follows:

$$i_a = 10 \cos 40 = 7.66$$

$$i_b = 10 \cos(40 - 120°) = 1.74$$

$$i_c = 10 \cos(40 - 240°) = -9.4$$

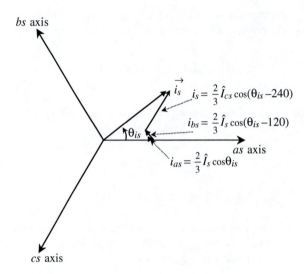

FIGURE 5.12 Space vector and its components in *abc* reference frame.

Calculate the resultant space vector.

Solution

The space vector at $\omega t = 40°$ is

$$\vec{i}_s = i_a(t) + i_b(t) \angle 120 + i_c(t) \angle 240$$

$$= 7.66 + 1.74 \angle 120 + (-9.4) \angle 240$$

$$= \frac{3}{2}(7.66 + j6.43)$$

$$= \frac{3}{2} \cdot 10 \angle 40$$

The projection of $\vec{i}_s$ on the Phase-a axis is 15 cos 40 = 11.49, which is 3/2 times i_a.

EXAMPLE 5.2

(a) The phase voltage magnitudes of a three-phase AC machine at time $\omega t = 0$ are $v_a = 240$ V, $v_b = -120$ V, and $v_c = -120$ V. Calculate the resulting space vector voltage.

(b) Recalculate the space vector at a different time, when $v_a = 207.8$ V, $v_b = 0$ V, and $v_c = -207.8$ V.

(c) Plot the space vector distribution in the air gap in the two cases.

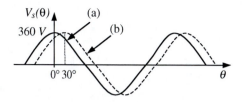

FIGURE 5.13 Plot of space vectors for Example 5.2 (a) and (b).

Solution

(a) The resulting space vector using Equation 5.13 is

$$\vec{v}_S(t) = v_a(t) + v_b(t) \angle 120 + v_c(t) \angle 240$$

$$\Rightarrow \vec{v}_S(t_0) = 240 + (-120)(\cos 120° + j \sin 120°) + (-120)(\cos 240° + j \sin 240°)$$

$$= 360 \angle 0° V$$

(b) The resulting space vector using Equation 5.13 is

$$\vec{v}_S(t) = v_a(t) + v_b(t) \angle 120 + v_c(t) \angle 240$$

$$\Rightarrow \vec{v}_S(t_1) = 207.8 + (0)(\cos 120° + j \sin 120°) + (-207.8)(\cos 240° + j \sin 240°)$$

$$= 360 \angle 30° V$$

(c) The plot is shown in Figure 5.13.

EXAMPLE 5.3

The phase voltage magnitudes of a three-phase AC machine at time t_0 are $v_a = 240$ V, $v_b = 50$ V, and $v_c = -240$ V. Calculate the resulting space vector voltage, and plot the space vector distribution in the air gap.

Solution

The resulting space vector using Equation 5.13 is as follows (Figure 5.14):

$$\vec{v}_S(t) = v_a(t) + v_b(t) \angle 120 + v_c(t) \angle 240$$

$$\Rightarrow \vec{v}_S(t_0) = 240 + (50)(\cos 120° + j \sin 120°) + (-240)(\cos 240° + j \sin 240°)$$

$$= 418.69 \angle 36.86° V$$

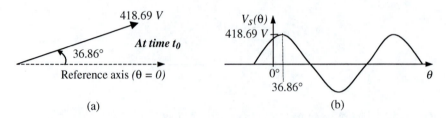

FIGURE 5.14 (a) Resultant voltage space vector. (b) Voltage space vector sinusoidal distribution in the air gap at t_0.

The voltages given in Example 5.2 are a balanced set at two different times, which corresponds to $\omega t = 0$ and $\omega t = 30°$ for parts (a) and (b), respectively. The peak magnitude of the resulting voltage space vector remains the same in the two cases, and the locations of these peaks along the machine axes are at $\theta = 0$ and $\theta = 30°$, which corresponds to the time dependence of the voltages. This is not coincidental, but will be so for a balanced set of voltages. In Example 5.3 the voltages are unbalanced, and the magnitude of the space vector depends on instantaneous values of the phase voltages.

5.4.4.1 Interpretation of Space Vectors

The space vectors, through one convenient and compact vector form, express the same resultant effect that the three individual phase variables would produce. For example, the stator mmf distribution in the air gap is a result of three phase currents i_a, i_b, and i_c, while the equivalent space vector current $\vec{i}_s(t)$ is developed in such a way that this imaginary current flowing through an equivalent winding of N_s turns would produce the same resultant mmf distribution.

The relationships between electrical and magnetic quantities are conveniently expressed with the help of space vectors. Using Equations 5.11, 5.12, and 5.14, we can write

$$\vec{F}_s(t) = \frac{N_s}{2}\vec{i}_s(t)$$

(5.15)

The mmf and current vector magnitudes are related by the scalar constant $N_s/2$, and they have the same angular orientation.

The flux density can be similarly shown to be

$$\vec{B}_s(t) = \frac{\mu_0 N_s}{2 l_g}\vec{i}_s(t)$$

(5.16)

5.4.4.2 Inverse Relations

Phase quantities can be derived from space vectors through inverse relations established using complex variable mathematics. We know that

$$|A|\angle\theta = |A|\cos\theta + j|A|\sin\theta$$

Applying this to Equation 5.13, we get

$$\vec{f}_{abc}(t) = f_a(t) - \frac{1}{2}(f_b(t) + f_c(t)) + j\frac{\sqrt{3}}{2}(f_b(t) - f_c(t)) = \frac{3}{2}f_a(t) + j\frac{\sqrt{3}}{2}(f_b(t) - f_c(t))$$

because $f_a(t) + f_b(t) + f_c(t) = 0$ for balanced three-phase systems and for circuits without a neutral connection. Therefore, the inverse relation for a Phase-a variable is

$$f_a(t) = \frac{2}{3}\mathrm{Re}\left[\vec{f}_{abc}(t)\right] \tag{5.17}$$

Similarly, it can be shown that

$$f_b(t) = \frac{2}{3}\mathrm{Re}\left[\vec{f}_{abc}(t)\angle 240°\right] \tag{5.18}$$

and

$$f_c(t) = \frac{2}{3}\mathrm{Re}\left[\vec{f}_{abc}(t)\angle 120°\right] \tag{5.19}$$

5.4.4.3 Resultant mmf in a Balanced System

In the typical operation of an AC machine, stator windings are supplied with balanced set voltages, and because the windings are electrically symmetrical, a balanced set of currents flows through the windings. Let us assume that the rotor is open circuited, and all the current flowing through the stator winding is the magnetizing current required to establish the stator mmf. The three phase currents have the same magnitude and frequency, but are 120° shifted in time with respect to each other. The currents in the time domain can be expressed as

$$i_a(t) = \hat{I}_M \cos\omega t$$
$$i_b(t) = \hat{I}_M \cos(\omega t - 120°) \tag{5.20}$$
$$i_c(t) = \hat{I}_M \cos(\omega t - 240°)$$

The space vector for the above balanced set of currents is as follows (using Equation 5.13):

$$\vec{i}_M(t) = \frac{3}{2}\hat{I}_M \angle\omega t \tag{5.21}$$

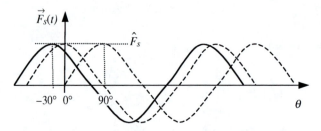

FIGURE 5.15 Resultant mmf space vector for $\omega t = -30°$, $0°$, and $90°$.

The resultant stator mmf space vector is as follows (Figure 5.15):

$$\vec{F}_S(t) = \frac{N_S}{2}\vec{i}_M(t) = \frac{3}{2}\frac{N_S}{2}\hat{I}_M \angle \omega t = \hat{F}_S \angle \omega t \tag{5.22}$$

The result shows that the stator mmf has a constant peak amplitude (because N_S and I_M are constants) that rotates around the stator circumference at a constant speed equal to the applied angular speed of the applied stator voltages. This speed is known as the *synchronous speed*. Unlike the single-phase stator mmf (shown in Figure 5.10b), the peak of the stator mmf resulting in the three-phase AC machine is rotating synchronously along the stator circumference, with the peak always located at $\theta = \omega t$. The mmf peak position is time-varying for the three-phase winding, whereas the peak mmf position for the single-phase winding is not time varying. The mmf wave is a sinusoidal function of the space angle θ. The wave has a constant amplitude and a space-angle ωt, which is a linear function of time. The angle ωt provides rotation of the entire wave around the air gap at a constant angular velocity ω. Thus, at a fixed time t_x, the wave is a sinusoid in space, with its positive peak displaced ωt_x from the reference $\theta = 0$. The polyphase windings excited by balanced polyphase currents produce the same general effect as that of spinning a permanent magnet about an axis perpendicular to the magnet, or as in the rotation of the DC-excited field poles.

The three-phase stator mmf is known as the rotating mmf, which can be equivalently viewed as a magnet rotating around the stator circumference at a constant speed. Note that with the vector sum of $F_a(\theta_e)$, $F_b(\theta_e)$, and $F_c(\theta_e)$ as described in Equations 5.9 and 5.10, with $i_a(t)$, $i_b(t)$, and $i_c(t)$ replaced by the balanced set of Equation 5.20, we will arrive at the same result.

EXERCISE 5.1

Show that

$$F_a(t) + F_b(t) + F_c(t) = \frac{3}{2}\frac{N_S}{2}\hat{I}_M \angle \omega t$$

with

$$i_a(t) = \hat{I}_M \cos \omega t$$

$$i_b(t) = \hat{I}_M \cos(\omega t - 120°)$$

$$i_c(t) = \hat{I}_M \cos(\omega t - 240°)$$

5.4.4.4 Mutual Inductance L_m and Induced Stator Voltage

In an ideal situation, the equivalent electrical circuit for the stator windings with no rotor existing consists of the applied stator voltage source and a set of windings represented by an inductance known as magnetizing or mutual inductance. The practical circuit extends on this ideal circuit by adding the stator winding resistance and the stator leakage inductance in series with the magnetizing inductance. The magnetizing inductance for the three-phase AC machine, including the effects of mutual coupling among the three phases, can be shown to be[4,5]

$$L_m = \frac{3}{2}\left[\frac{\pi\mu_0 rl}{l_g}\left(\frac{N_S}{2}\right)^2\right] \tag{5.23}$$

where r is the radius to the air gap, l is the rotor axial length, and l_g is the air gap length. Note that the form of Equation 5.23 is the same as that of a simple inductor given by $L = N^2/\Re$, where N is the number of turns, and $\Re = Reluctance = flux - path\ length/\mu \times cross\text{-}sectional\ area$. The voltage induced in the stator windings due to the magnetizing current o wing through L_m in space vector form is

$$\vec{e}_{ms}(t) = j\omega L_m \vec{i}_M(t) \tag{5.24}$$

The magnetizing ux-density established by the magnetizing current is as follows (from Equation 5.16):

$$\vec{B}_{ms}(t) = \frac{\mu_0 N_S}{2l_g}\vec{i}_M(t)$$

which gives

$$\vec{i}_M(t) = \frac{2l_g}{\mu_0 N_S}\vec{B}_{ms}(t) \tag{5.25}$$

Using the expression for L_m from Equation 5.23 and the expression for $\vec{i}_M(t)$ in terms of $\vec{B}_{ms}(t)$, the induced voltage is

$$\vec{e}_{ms}(t) = j\omega \frac{3}{2} \pi rl \frac{N_S}{2} \vec{B}_{ms}(t)$$

The induced voltage can be interpreted as the back-emf induced by the flux-density B_{ms}, which is rotating at the synchronous speed. For a P-pole machine, the expression for the induced voltage is

$$\vec{e}_{ms}(t) = j\omega \frac{3}{2} \pi rl \frac{N_S}{P} \vec{B}_{ms}(t) \tag{5.26}$$

5.4.5 Types of AC Machines

The second rotating mmf needed for torque production in AC machines is established by the rotor circuit. The interaction of the two rotating mmfs, essentially chasing each other at synchronous speed, is what produces torque. The method through which the rotor mmf is established differentiates the types of AC machines. Broadly, AC machines can be classified into two categories, synchronous machines and asynchronous machines. In synchronous machines, the rotor always rotates at synchronous speed. The rotor mmf is established by using a permanent magnet or an electromagnet created by feeding DC currents in a rotor coil. The latter types of synchronous machines are typically large machines used in electric power-generating systems. PM machines are more suitable for EV and HEV applications, because these offer higher power density and superior performance compared to the rotor-fed synchronous machines. The several types of permanent magnet AC machines will be discussed in Chapter 6. The rotor-fed synchronous machines will not be discussed further in this book, because these are not of interest for EV and HEV applications. In the asynchronous type AC machine, the rotor rotates at a speed that is different but close to the synchronous speed. These machines are known as induction machines, which in the more common configurations, are fed only from the stator. The voltages in the rotor circuit are induced from the stator, which in turn, induce the rotor rotating mmf and, hence, the name induction machines. Induction machines generally labeled as AC machines will be discussed in the following section.

5.5 INDUCTION MACHINES

The two types of induction machines are the squirrel cage induction machines and the wound rotor induction machines. Squirrel cage induction machines are the workhorses of the industry because of their rugged construction and low cost. The rotor windings consist of short-circuited copper or aluminum bars that form the shape of a squirrel cage. The squirrel cage of an induction motor is shown in Figure 5.16. The rotor winding terminals of the wound rotor induction machines are brought

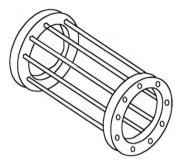

FIGURE 5.16 The squirrel cage of an induction motor.

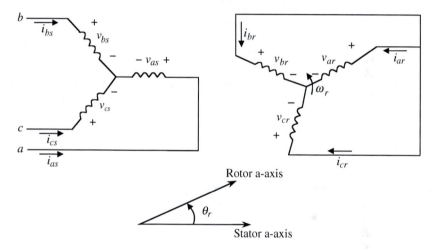

FIGURE 5.17 Stator and rotor electric circuit and magnetic axes of a three-phase induction machine.

outside with the help of slip rings for external connections, which are used for speed control. Squirrel cage induction motors are of greater interest for EV and HEV and most other general purpose applications and, hence, will be discussed further.

The stator windings of induction machines are the same as discussed in the previous section. The rotor, usually made of stacked laminations, has copper or aluminum rotor bars molded around the periphery in axial direction. The bars are short-circuited at the ends through electrically conducting end rings. The electrical equivalent circuit of a three-phase induction machine along with the direction of Phase-a stator and rotor magnetic axes are shown in Figure 5.17. Rotor windings have been short-circuited, and the angle between the rotor and stator axes is θ_r, which is the integral of the rotor speed ω_r.

When a balanced set of voltages is applied to the stator windings, a magnetic field is established, which rotates at synchronous speed, as described in Section 5.4. By Faraday's Law ($e = Blv$), as long as the rotor rotates at a speed other than the synchronous speed (even at zero rotor speed), the rotor conductors are cutting the stator magnetic field, and there is a rate of change of flux in the rotor circuit, which will induce a voltage in the rotor bars. This is also analogous to transformer action, where a time-varying AC flux established by the primary winding induces voltage in the secondary set of windings. The induced voltage will cause rotor currents to flow in the rotor circuit, because the rotor windings or bars are short-circuited in the induction machine. The induction machine can be thought of as a transformer with short-circuited secondary, i.e., rotor windings. The rotor-induced voltages and the current have a sinusoidal space distribution, because these are created by the sinusoidally varying (space sinusoids) stator magnetic field. The resultant effect of the rotor bar currents is to produce a sinusoidally distributed rotor mmf acting on the air gap.

The difference between the rotor speed and the stator synchronous speed is the speed by which the rotor is slipping from the stator magnetic field, and it is known as the slip speed:

$$\omega_{slip} = \omega_e - \omega_m \tag{5.27}$$

where ω_e is the synchronous speed, and ω_m is the motor or rotor speed. The slip speed expressed as a fraction of the synchronous speed is known as the slip:

$$s = \frac{\omega_e - \omega_m}{\omega_e} \tag{5.28}$$

Rotor bar voltages, current, and magnetic field are of the slip speed or slip frequency with respect to the rotor. The slip frequency is given by

$$f_{slip} = \frac{\omega_e - \omega_m}{2\pi} = sf, \quad \text{where} \quad f = \frac{\omega_e}{2\pi} \tag{5.29}$$

From the stator perspective, the rotor voltages, currents, and rotor mmf have synchronous frequency, because the rotor speed of ω_m is superimposed on the rotor variables' speed of ω_{slip}.

5.5.1 PER-PHASE EQUIVALENT CIRCUIT

Steady state analysis of induction motors is often carried out using the per-phase equivalent circuit. A single-phase equivalent circuit is used for the three-phase induction machine, assuming a balanced set as shown in Figure 5.18. The per-phase equivalent circuit consists of the stator loop and the rotor loop, with the magnetic circuit parameters in the middle. The inductance representing the magnetization current path is in the middle of the circuit, along with an equivalent core loss resistance. For the stator and rotor electrical parameters, the circuit includes the stator winding resistance and leakage reactance and the rotor winding resistance and leakage reactance. A slip-dependent equivalent resistance represents the mechanical power delivered at the shaft due to the energy conversion in the air gap coupled

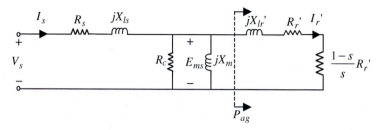

FIGURE 5.18 Steady state per-phase equivalent circuit of an induction motor.

electromagnetic circuit. The electrical input power supplied at the stator terminals converts to magnetic power and crosses the air gap. The air gap power P_{ag} is converted to mechanical power delivered at the shaft after overcoming losses in the rotor circuit.

Although the per-phase equivalent circuit is not enough to develop controllers that demand good dynamic performance like in an EV or HEV, the circuit provides a basic understanding of induction machines. The vast majority of applications of induction motors are for adjustable speed drives, where controllers designed for good steady state performance are adequate. The circuit allows the analysis of a number of steady state performance features. The parameters of the circuit model are as follows:

E_{ms} = Stator-induced emf per phase
V_s = Stator terminal voltage per phase
I_s = Stator terminal current
R_s = Stator resistance per phase
X_{ls} = Stator leakage reactance
X_m = Magnetizing reactance
X_{lr}' = Rotor leakage reactance referred to stator
R_r' = Rotor resistance referred to stator
I_r' = Rotor current per phase referred to stator

Note that the voltages and currents described here in relation to the per-phase equivalent circuit are phasors and not space vectors. The power and torque relations are

$$P_{ag} = \text{Air gap power} = 3\left|I_r'\right|^2 \frac{R_r'}{s}$$

$$P_{dev} = \text{Developed mechanical power} = 3\left|I_r'\right|^2 \frac{(1-s)R_r'}{s}$$

$$= (1-s)P_{ag}$$

$$= T_e\omega_m$$

$$P_R = \text{Rotor copper loss} = 3\left|I_r'\right|^2 R_r'$$

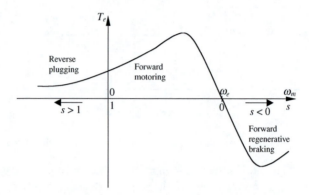

FIGURE 5.19 Steady-state torque–speed characteristics of an induction motor.

The electromagnetic torque is given by

$$T_e = 3\left|I_r'\right|^2 \frac{(1-s)R_r'}{s\omega_m}$$

$$= \frac{3R_r'}{s\omega_s} \frac{V_s^2}{\left(R_s + R_r'/s\right)^2 + \left(X_s + X_r'\right)^2}$$

(5.30)

The steady-state torque–speed characteristics of the machine are as shown in Figure 5.19. The torque produced by the motor depends on the slip and the stator currents, among other variables. The induction motor starting torque, while depending on the design, is lower than the peak torque achievable from the motor. The motor is always operated in the linear region of the torque–speed curve to avoid the higher losses associated with high slip operation. In other words, operating the machine at small slip values maximizes the efficiency.

The value of the rotor circuit resistance determines the speed at which the maximum torque will occur. In general, the starting torque is low, and the maximum torque occurs close to the synchronous speed, when the slip is small. The motor draws a large current during line starting from a fixed AC source, which gradually subsides as the motor reaches the steady state speed. If the load requires a high starting torque, the motor will accelerate slowly. This will make a large current flow for a longer time, thereby creating a heating problem.

Nonlinearity at speeds below the rated condition is due to the effects of leakage reactances. At higher slip values, the frequencies of the rotor variables are large, resulting in dominating impedance effects from rotor leakage inductance. The air gap flux cannot be maintained at the rated level under this condition. Also, large values of rotor current (which flows at high slip values) cause a significant voltage drop across the stator winding leakage impedance $(R_s + j\omega L_{ls})$, which reduces the induced voltage and, in turn, the stator mmf flux density $\hat{B}_{ms}$.

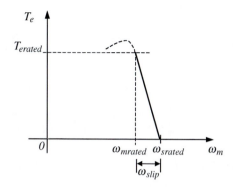

FIGURE 5.20 Torque–speed characteristics of an induction motor for rated flux condition.

5.5.2 SIMPLIFIED TORQUE EXPRESSION

A simplified linear torque expression is sufficient to analyze the motor–load inter-action, because the induction motor is invariably always operated in the linear region at maximum flux density $\hat{B}_{ms}$, with the help of a power electronic feed circuit. The segment of interest in the torque speed characteristic curve of the induction motor is shown by the firm line in Figure 5.20. Synchronous speed is set by the applied voltage frequency, and the slope of the linear region is set by design dimensions and material properties. Hence, assuming that the stator flux density is kept constant at its rated value, steady state torque can be expressed linearly as a function of slip:

$$T_e = K_{IM}\omega_{slip}$$

where K_{IM} is a constant. In order to find out what this constant depends on, one needs to trace back to the origin of torque production in the machine. Electromagnetic torque is produced by the tendency of the stator and rotor mmfs to align with each other. Rotor mmf is due to rotor current. The principle of torque production essentially lies in the Lorenz force law ($F = Bil$). Therefore, the electromagnetic torque produced in an induction machine at steady state can be expressed as

$$T_e = k_M \hat{B}_{ms} \hat{I}_r' \qquad (5.31)$$

where k_M is a machine constant, $\hat{B}_{ms}$ is the equivalent peak stator mmf flux density for the three-phase machine, and $\hat{I}_r$ is the peak equivalent rotor current. Note that these are different from the single-phase equivalent circuit per phase quantities. Mohan[5] showed that this machine constant is

$$k_M = \pi r l \frac{N_S}{2}$$

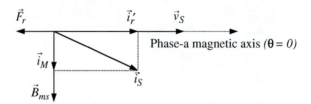

FIGURE 5.21 Space vector diagram of stator and rotor variables at $t = 0$.

where r is the radius to the air gap, l is axial length of the machine, and N_S is the equivalent number of turns.

In order to find a relation between $\hat{B}_{ms}$ and rotor current $\hat{I}_r$, let us denote the rotor mmf by the space vector $\vec{F}_r(t)$. The stator windings must carry currents in addition to the magnetizing current $\vec{i}_M(t)$ to support the currents induced in the rotor by transformer action to create $\vec{F}_r(t)$. These rotor currents referred to stator, or in other words, the additional stator current is represented by $\vec{i}_r'(t)$ (with magnitude $\hat{I}_r$) and is related to $\vec{F}_r(t)$ by

$$\vec{i}_r'(t) = \frac{\vec{F}_r}{N_S/2}$$

The total stator current is the sum of magnetizing current and referred rotor current:

$$\vec{i}_S(t) = \vec{i}_M(t) + \vec{i}_r'(t) \tag{5.32}$$

These space vectors are shown in Figure 5.21. Rotor leakage inductance L_{lr}' has been neglected in this diagram for simplification. Although this is an idealistic situation, it is an important assumption that helps one grasp the basic concepts of torque production in induction machines. Neglecting the rotor leakage reactance is equivalent to assuming that all of the flux created by the rotor bar currents crosses the air gap and links the stator windings, and that there are no leakage fluxes in the rotor. The rated speed on induction machines is close to the synchronous speed, and the machine usually operates near the rated condition with a small slip. At small slip values, the slip speed ω_{slip} is small, and it is justified to neglect the effect of rotor leakage inductance, which is small anyway. Mohan[5] showed that under this simplifying assumption, rotor bar currents induced by Faraday's law are proportional to stator flux density and slip speed, the peak of which can be denoted by

$$\hat{I}_r' = k_r \hat{B}_{ms} \omega_{slip} \tag{5.33}$$

where k_r is a machine design constant. Substituting $\hat{I}_r'$ from Equation 5.33 in the torque equation,

$$T_e = k_m k_r \hat{B}_{ms}{}^2 \omega_{slip}$$

The electromagnetic torque for fixed stator flux density is

$$T_e = K_{IM} \omega_{slip} \qquad (5.34)$$

where $K_{IM} = k_m k_r \hat{B}_{ms}{}^2$.

The simple torque expression presents a convenient method of defining the torque–speed relation of an induction machine linearly near the rated operating point, similar to the DC machine relationship. The expression can be used to find the steady state operating point of an induction-machine-driven EV or HEV by finding the point of intersection of the machine torque–speed characteristics and the road load force–speed characteristics. When the motor rotates at synchronous speed, the slip speed is zero, and the motor does not produce torque. In practice, the machine never reaches the synchronous speed, even in an unloaded condition, because a small electromagnetic torque is needed to overcome the no-load losses that include friction and windage losses. Slip speed is small up to the rated torque of the machine, and hence, it is reasonable to neglect rotor leakage inductance, which gives the linear torque–speed relationship. The machine runs close to the synchronous speed under no-load condition with a small slip. As the machine is loaded from a no-load condition, the slip starts to increase, and the speed approaches the rated speed condition. Beyond the rated condition, the machine operates with a higher slip, and the assumption of neglecting the leakage inductance starts to fall apart. Hence, this portion of torque–speed characteristics is shown by a dotted curve in Figure 5.20.

Induction motors for EVs and HEVs and other high-performance applications are supplied from a variable voltage, variable frequency AC source. Varying the frequency changes the rated flux and synchronous speed of the machine, which essentially causes the linear torque–speed curve of Figure 5.20 to move horizontally along the speed axis toward the origin.

EXAMPLE 5.4

The vehicle road load characteristic on a level road is $T_{TR} = 24.7 + 0.0051\, \omega_{wh}{}^2$. The induction motor torque–speed relationship in the linear region is given by $T_e = K_{IM}(40 - \omega_m)$, including the gear ratio of the transmission system. The rated torque of 40 Nm is available at a speed of 35 rad/s. Find the steady state operating point of the vehicle.

Solution

The induction motor torque constant is

$$K_{IM} = 40/40 - 35 = 8\,NM/\text{rads/s}$$

The steady state operating point is obtained by solving the vehicle road load characteristic and the motor torque speed characteristic, which gives

$$\omega^* = 36.08 \text{ rads/s} \quad \text{and} \quad T^* = 31.34\,Nm$$

5.5.3 SPEED CONTROL METHODS

The speed of an induction motor can be controlled in two ways, by varying stator terminal voltage and stator frequency.

Changing the terminal voltage changes the torque output of the machine, as is evident from Equation 5.30. Note that changing the applied voltage does not change the slip for maximum torque. The speed control through changing the applied frequency is based on the frequency and synchronous speed relation $\omega_e = 4\pi f/P$; changing f changes ω_e. Figures 5.22 and 5.23 show variations in torque–speed characteristics with changes in voltage and frequency, respectively. What is needed to drive the induction motor is a power electronics converter that will convert the available constant voltage into a variable voltage, variable frequency output, according to the command torque and speed. The top-level block diagram of such a drive system is shown in Figure 5.24. First-generation controllers of induction motor drives used in EVs employed slip control (constant V/Hz control) using a table of slip vs. torque. The performance of such a drive for vehicle applications is poor, because the concept of V/Hz control is based on a steady state equivalent circuit of the machine. The dynamic performance of the machine improves significantly using vector control. The dq-axes transformation theory for induction motors will be presented as an introduction to vector control theory in the next section, after the discussion on regenerative braking. Induction motor drives will be discussed in Chapter 8.

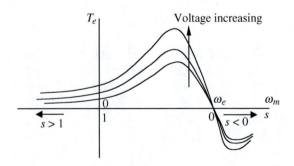

FIGURE 5.22 Torque–speed profile at different voltages with fixed supply frequency.

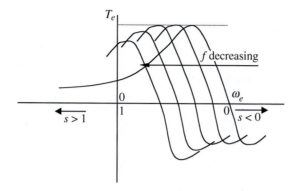

FIGURE 5.23 Torque–speed profile with variable frequency but constant V/f ratio.

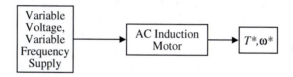

FIGURE 5.24 Induction motor drive.

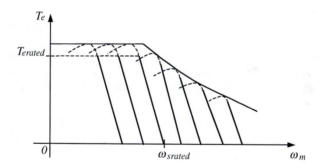

FIGURE 5.25 Torque–speed operating envelope for the induction motor.

Figure 5.25 shows the envelope of the torque speed characteristics of an induction motor. Using a power electronics controlled drive, it is possible to achieve constant power characteristics from an induction motor at higher speeds, a feature that is so important for EV and hybrid vehicle motor drives.

5.6 REGENERATIVE BRAKING

One of the advantages of using electric motors for vehicle propulsion is saving energy during vehicle braking through regeneration. The regenerated energy can be used to recharge the batteries of an EV or HEV. It is important to note that it will not be possible to capture all of the energy available during vehicle braking, especially when sudden stops are commanded. The energy available during braking is the kinetic energy that was acquired by the vehicle during acceleration. The energy is typically too high to be processed by the electric motor used for propulsion. Processing high energy in a relatively short time would require a huge motor or, in other words, a motor with high power ratings, which is impractical. Hence, EVs and HEVs must be equipped with the mechanical brake system, even though the electric motor drive is designed with regeneration capability. However, regeneration can save a significant portion of energy, extending the range of a vehicle. The vehicle supervisory controller decides the amount of braking needed from the mechanical system based on the braking command of the driver, the amount of regeneration possible, and vehicle velocity.

In the regenerative braking mode, the kinetic energy of the vehicle is processed by the electric machine and returned to the energy source. From the machine perspective, this is no different than operating the machine in the generator mode. The electric machine converts the mechanical power available from vehicle kinetic energy and converts it to electrical energy — the flow of energy is from the wheels to the source. Regenerative braking can increase the range of EVs by about 10 to 15%.

The induction machine works as a generator when it is operated with a negative slip, i.e., the synchronous speed is less than the motor speed $\omega_m > \omega_e$. Negative slip makes the electromagnetic torque negative during regeneration or the generating mode. In the negative slip mode of operation, voltages and currents induced in the rotor bars are of opposite polarity compared to those in the positive slip mode. Electromagnetic torque acts on the rotor to oppose the rotor rotation, thereby decelerating the vehicle.

The motor drives for EVs are always four-quadrant drives, meaning that the electric motor is controlled by the drive to deliver positive or negative torque at positive or negative speed. The transition from forward motoring to regeneration can be explained with the help of Figure 5.26 for four-quadrant induction motor drives. The linear segments of the induction motor torque–speed curves for several operating frequencies are shown in the figure. Consider the frequencies f_1 and f_2.

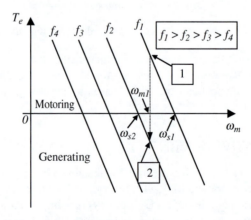

FIGURE 5.26 Transition from motoring to generating using a four-quadrant drive.

The curves are extended in the negative torque region to show the characteristics during regeneration. Suppose initially, the electric vehicle is moving forward, being driven by the positive torque delivered by the induction motor, and the steady state operating point in this condition is at Point 1. Now, the vehicle driver presses the brake to slow the vehicle. The vehicle system controller immediately changes the motor drive frequency to f_2 such that $\omega_{s2} < \omega_{m1}$. The operating point shifts to Point 2 immediately, because the motor speed cannot change instantaneously due to inertia of the system. At Point 2, the slip and the electromagnetic torque are negative, and the motor is regenerating. The vehicle will slow from this condition onward. As the motor speed decreases and falls below the synchronous speed, the operating frequency needs to be changed to a lower value so that generating mode operation can be maintained. The power electronics drive is responsible for establishing the shifted linear torque–speed curves with different synchronous speeds for the induction machine at different frequencies, as shown. The drive circuit does so by changing the frequency of the supply voltage. The regenerative braking mode continues as long as there is kinetic energy available and the driver wishes to slow the vehicle. Similar to starting, regeneration has to be achieved in a controlled way, so that the power rating of the machine is not exceeded. The amount of kinetic energy to be converted within the desired stopping time determines the power that is to be handled by the machine.

5.7 *dq* MODELING

The *dq* modeling relates to the transformation of three-phase variables in the *abc* coordinate system into an equivalent two-phase coordinate system that has an arbitrary speed in a given reference frame.[6,7] In the *dq* coordinate system, the *d*-axis is along the direct magnetic axis of the resultant mmf, while the *q*-axis is in quadrature to the direct axis. The *dq* modeling of AC machines enables the development of an electric machine controller, which operates in the inner loop with respect to the outer loop of the system level controller. The *dq* modeling analysis provides all necessary transformation equations required to implement the inner-loop controller. Furthermore, the electromagnetic torque T_e expression in terms of machine variables (current, ux linkage, etc.) in the *dq* model is often used to estimate the torque for closed-loop control. These machine variables are easily measurable using sensors, which is much simpler than torque measurement. The simpli ed torque e xpression of Equation 5.34 or the steady state torque expression of Equation 5.30 are inadequate for dynamic controller implementation, as will be evident through later discussions.

Signi cant coupling e xists between the stator and rotor variables of a three-phase AC machine. The objective of *dq* modeling is to transform the three-phase *abc*-variables into *dq*-variables in a suitable reference frame, such that the coupling disappears. The space vector approach is retained in the *dq* reference frame, because these vectors help express the complex three-phase equations of AC machines in a compact form and also provide a simple relation for transformation between *abc* and *dq* reference frames. *q*-axes equivalent stator currents, designated as $i_{sd}(t)$ and $i_{sq}(t)$, establish the same stator mmf as that produced by the *a, b,* and *c* stator winding currents $i_a(t)$, $i_b(t)$, and $i_c(t)$. The d- and q-axes equivalent stator currents $i_{sd}(t)$ and $i_{sq}(t)$ flow through a fictitious orthogonal set of windings along the d- and q-axes. The stator mmf produced by the abc stator currents or the equivalent dq stator currents is given by Equation (5.15) and is repeated here for convenience:

$$\vec{F}_S(t) = \frac{N_S}{2}\vec{i}_S(t)$$

Several choices exist to define the relation between the abc-axes variables and the dq-axes variables, because the transformation is from a three-phase set of variables to a two-phase set of variables. One choice is to take the dq-variables to be 2/3 times the projection of $\vec{f}_{abc}$ on the d and q axes. The same space vector magnitude as the peak value of the individual phase time-phasor variables as we have seen earlier is given by $2/3 \cdot \vec{f}_{abc}$. Another possible choice is to take the dq-variables to be $\sqrt{2/3}$ times the projection of $\vec{f}_{abc}$ on the d and q axes. The $\sqrt{2/3}$ ratio between the dq-variables and the abc-variables conserves power without any multiplying factor in the dq and abc reference frames and, hence, is known as the power invariant transformation. In order for the equivalent dq windings to establish the same stator mmf $\vec{F}_S(t)$ as is done by the abc windings, the number of turns in the equivalent sinusoidally distributed orthogonal windings must be $2/3 N_S$ or $\sqrt{2/3} N_S$, depending the ratio chosen for transformation. We will arbitrarily choose the multiplying factor of 2/3 to define the transformation. In this case, the q- and d-axes variables are the projections of a, b, and c variables on the q and d axes, respectively multiplied by 2/3. Mathematically stated, for a general variable f (representing voltage, current, or flux linkage), the q- and d-axes variables in terms of the a, b, and c variables are

$$f_q = \frac{2}{3}\left[f_a \cos\theta + f_b \cos(\theta - 120°) + f_c \cos(\theta + 120°)\right]$$

$$f_d = \frac{2}{3}\left[f_a \sin\theta + f_b \sin(\theta - 120°) + f_c \sin(\theta + 120°)\right]$$

(5.35)

The transformation relation holds for stator as well as for rotor variables. A third variable is required to obtain a unique transformation, which comes from the neutral terminal. Representing the neutral terminal variable as a zero-sequence component, we have

$$f_0 = \frac{1}{3}\left[f_a + f_b + f_c\right]$$

(5.36)

All of the three-phase systems we will consider here are balanced. Furthermore, the AC machine windings are connected either in Δ or Y without a neutral connection. Therefore, for all practical purposes,

$$f_a + f_b + f_c = 0 \Rightarrow f_0 = 0$$

Henceforth, we will concentrate only on the d and q variables and ignore the zero-sequence component.

The variables in the dq reference frame can be expressed in the space vector form as

$$\vec{f}_{qd}(t) = f_q(t) - jf_d(t) = \frac{2}{3}e^{-j\theta}\left[f_a(t) + e^{j\frac{2\pi}{3}}f_b(t) + e^{-j\frac{2\pi}{3}}f_b(t) \right]$$

$$= \frac{2}{3}e^{-j\theta}\vec{f}_{abc}(t)$$

(5.37)

The result shows that the dq space vector $\vec{f}_{abc}(t)$ is 2/3 of the space vector $\vec{f}_{abc}(t)$ along the d- and q-axes. Equating the real and imaginary parts of Equation 5.37, the transformation matrix between the abc and dq variables is as follows:

$$\begin{bmatrix} f_d(t) \\ f_q(t) \\ 0 \end{bmatrix} = T_{abc->dq}\begin{bmatrix} f_a(t) \\ f_b(t) \\ f_c(t) \end{bmatrix}$$

(5.38)

where

$$T_{abc->dq} = \frac{2}{3}\begin{bmatrix} \cos(\theta) & \cos(\theta - 2\pi/3) & \cos(\theta + 2\pi/3) \\ \sin(\theta) & \sin(\theta - 2\pi/3) & \sin(\theta + 2\pi/3) \\ 0.5 & 0.5 & 0.5 \end{bmatrix}$$

This transformation is known as Park's transformation. The abc variables are obtained from the dq variables through the inverse of the Park transform:

$$T_{dq->abc} = \begin{bmatrix} \cos(\theta) & \sin(\theta) & 1 \\ \cos(\theta - 2\pi/3) & \sin(\theta - 2\pi/3) & 1 \\ \cos(\theta + 2\pi/3) & \sin(\theta + 2\pi/3) & 1 \end{bmatrix}$$

(5.39)

The projections from the abc frame and the dq frame to form the same space vector for a three-phase AC machine current variable are shown in Figure 5.27. The orientation of the dq-axes with respect to the abc-axes is at an arbitrary angle θ. The direct-axis and quadrature-axis stator currents i_{ds} and i_{qs} are a set of fictitious two-phase current components, and the same is true for other dq variables. The dq-reference frame can be stationary with respect to the stator or rotating at an arbitrary speed, such as at rotor speed or at synchronous speed. Again, when a stationary dq reference frame is used, the dq-axes may be at any arbitrary angle with respect to our chosen reference phase a-axis.

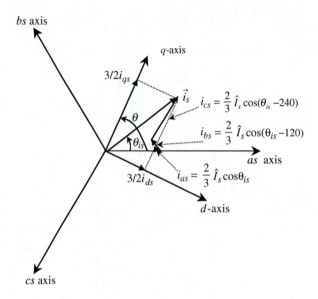

FIGURE 5.27 Transformation from three-phase variables to *dq*-axes variables.

5.7.1 ROTATING REFERENCE FRAME

Although the speed of the *dq*-winding can be arbitrary as mentioned previously, three of those are the most suitable for machine analysis. These three speeds of *dq* windings are 0, ω_m, and ω_e. The zero speed is known as the stationary reference frame, where, typically, the stationary *d*-axis is aligned with the phase *a*-axis of the stator. The angle of transformation θ in Equations 5.37, 5.38, and 5.39 is 0 in this case. The *d*- and *q*-axes variables oscillate at synchronous frequency in the balanced sinusoidal steady state. When ω_e is chosen as the speed of the reference *dq* frame, all the associated variables in the stator and in the rotor *dq* windings appear as DC variables in the balanced sinusoidal steady state. For an arbitrary speed of reference *dq* windings, the angle of transformation is

$$\theta = \int_0^t \omega(\xi)\,d\xi + \theta_0 \tag{5.40}$$

EXAMPLE 5.5

The three-phase currents in an AC machine are

$$i_a(t) = 10 \cos 377t$$

$$i_b(t) = 10 \cos\left(377t - \frac{2\pi}{3}\right)$$

$$i_c(t) = 10 \cos\left(377t + \frac{2\pi}{3}\right)$$

Calculate the currents as *dq* variables:

(a) Stationary reference frame
(b) Synchronous reference frame

Solution

(a) Using Equation 5.38 and $\theta = 0$, the *dq* current variables in the stationary reference frame are

$$i_d^s(t) = -10\sin(377t)$$
$$i_q^s(t) = 10\cos(377t)$$

(b) Using Equation 5.38 and $\theta = \omega t = 377t$ with $\theta_0 = 0$, the *dq* current variables in the synchronously rotating reference frame are

$$i_d^e(t) = 0$$
$$i_q^e(t) = 10$$

5.7.2 INDUCTION MACHINE *dq* MODEL

Let us assume that the stator and rotor voltages and currents in the three-phase model vary arbitrarily in time. The voltage balance equations in the stator and rotor circuits (shown in Figure 5.17) of the three-phase induction machine in space vector form are

$$\vec{v}_{abcs} = R_s\vec{i}_{abcs} + p\vec{\lambda}_{abcs}$$
$$\vec{v}_{abcr} = R_r\vec{i}_{abcr} + p\vec{\lambda}_{abcr}$$

The rotor winding voltage $\vec{v}_{abc}$ is zero for a squirrel cage induction machine with short-circuited rotor windings but will be represented as such for generality. Rotor variables are represented here without the rotor-to-stator referral symbol for the sake of simplicity, but it is to be understood to be implicitly incorporated in the equations. The stator and rotor ux linkages include coupling ef ects between the windings of stator and rotor circuits as well as between stator and rotor windings. Accounting for all magnetic coupling and assuming magnetic linearity ($\lambda = Li$), the phase-variable form of the voltage equations in the *abc* frame can be derived as

$$\vec{v}_{abcs} = R_s\vec{i}_{abcs} + L_S\left(p\vec{i}_{abcs}\right) + L_m\left(p\vec{i}_{abcr}\right)e^{j\theta_r} + j\omega_r L_m\vec{i}_{abcr}e^{j\theta_r} \qquad (5.41)$$

$$\vec{v}_{abcr} = R_r\vec{i}_{abcr} + L_r\left(p\vec{i}_{abcr}\right) + L_m\left(p\vec{i}_{abcs}\right)e^{-j\theta_r} - j\omega_r L_m\vec{i}_{abcs}e^{-j\theta_r} \qquad (5.42)$$

where $\omega_r = p\theta_r = d\theta_r/dt$ is the rotor speed, and $L_S = L_{ls} + L_m$ and $L_r = L_{lr} + L_m$. L_{ls} and L_{lr} are the stator and rotor leakage inductances. The derivation can be carried out using the space vector approach[4] or reference frame transformation.[6]

Multiplying by $e^{j\theta}$ and applying dq transformation, Equations 5.41 and 5.42 can be transformed into a general reference frame rotating at a speed ω as

$$\vec{v}_{qds} = R_s\vec{i}_{qds} + L_S\left(p\vec{i}_{qds}\right) + L_m\left(p\vec{i}_{qdr}\right) + j\omega\left(L_S\vec{i}_{qds} + L_m\vec{i}_{qdr}\right) \tag{5.43}$$

$$\vec{v}_{qdr} = R_r\vec{i}_{qdr} + L_r\left(p\vec{i}_{qdr}\right) + L_m\left(p\vec{i}_{qds}\right) + j\left(\omega - \omega_r\right)\left(L_r\vec{i}_{qdr} + L_m\vec{i}_{qds}\right) \tag{5.44}$$

The matrix form of the induction motor model in the arbitrary dq reference frame is

$$
\begin{bmatrix} v_{ds} \\ v_{qs} \\ v_{dr} \\ v_{qr} \end{bmatrix} =
\begin{bmatrix}
R_S & -\omega L_S & 0 & -\omega L_m \\
\omega L_S & R_S & \omega L_m & 0 \\
0 & -(\omega - \omega_r)L_m & R_r & -(\omega - \omega_r)L_r \\
(\omega - \omega_r)L_m & 0 & (\omega - \omega_r)L_r & R_r
\end{bmatrix}
\begin{bmatrix} i_{ds} \\ i_{qs} \\ i_{dr} \\ i_{qr} \end{bmatrix}
$$
$$
+ \begin{bmatrix}
L_S & 0 & L_m & 0 \\
0 & L_S & 0 & L_m \\
L_m & 0 & L_r & 0 \\
0 & L_m & 0 & L_r
\end{bmatrix}
\begin{bmatrix} pi_{ds} \\ pi_{qs} \\ pi_{dr} \\ pi_{qr} \end{bmatrix} \tag{5.45}
$$

The dq-equivalent circuit model for the induction machine in circuit schematic form is shown in Figure 5.28.

5.7.3 POWER AND ELECTROMAGNETIC TORQUE

The power into the three-phase machine needs to be analyzed in terms of dq variables to arrive at the electromagnetic torque expression, which is used in the motor control loop. The power into the induction machine is the product of the phase voltage and phase currents, given as

$$P_{in} = \left(v_{as}i_{as} + v_{bs}i_{bs} + v_{cs}i_{cs}\right) + \left(v_{ar}i_{ar} + v_{br}i_{br} + v_{cr}i_{cr}\right) \tag{5.46}$$

The input power is embedded in the real part of the product of the voltage space vector and the current space vector conjugate of the stator and rotor variables. The real part can be calculated as follows:

$$\text{Re}\left(\vec{v}_{abc}\vec{i}_{abc}^*\right) = \text{Re}\left[\left(v_a + v_b\angle 120 + v_c\angle 240\right)\cdot\left(i_a + i_b\angle -120 + i_c\angle -240\right)\right]$$

$$= \frac{3}{2}\left(v_a i_a + v_b i_b + v_c i_c\right)$$

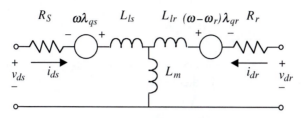

d-axis equivalent circuit

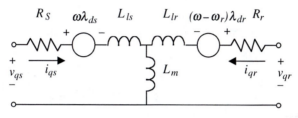

q-axis equivalent circuit

FIGURE 5.28 *d*- and *q*-axes circuits of the induction machine.

Using the above relation in the power input equation for the induction machine, we get

$$P_{in} = \frac{2}{3}\left[\text{Re}\left(\vec{v}_{abcs}\vec{i}_{abcs}^*\right) + \text{Re}\left(\vec{v}_{abcr}\vec{i}_{abcr}^*\right)\right]$$

Now, using Equation 5.37:

$$P_{in} = \frac{2}{3}\left[\text{Re}\left[\left(\frac{3}{2}e^{j\theta}\vec{v}_{qds}\right)\left(\frac{3}{2}e^{-j\theta}\vec{i}_{qds}^*\right)\right] + \text{Re}\left[\left(\frac{3}{2}e^{j\theta}\vec{v}_{qdr}\right)\left(\frac{3}{2}e^{-j\theta}\vec{i}_{qdr}^*\right)\right]\right]$$

$$= \frac{3}{2}\left[\text{Re}\left(\vec{v}_{qds}\vec{i}_{qds}^*\right) + \text{Re}\left(\vec{v}_{qdr}\vec{i}_{qdr}^*\right)\right]$$

(5.47)

The input power expression in scalar form is

$$P_{in} = \frac{3}{2}\left(v_{ds}i_{ds} + v_{qs}i_{qs} + v_{dr}i_{dr} + v_{qr}i_{qr}\right)$$

(5.48)

The multiplying factor 3/2 is due to our choice of the factor 2/3 for the ratio of *dq* and *abc* variables. The input power expression of Equation 5.48 would not have this 3/2 factor, had we chosen $\sqrt{2/3}$ as the proportionality constant between the *dq* and *abc* variables.

Upon expanding the right-hand side of Equation 5.48 and extracting the output electromechanical power P_e,[4] we get

$$P_e = \frac{3}{2} \mathrm{Im}\left[\omega_r L_m \vec{i}_{qds} \vec{i}_{qdr}^*\right] \tag{5.49}$$

where ω_r is the rotor angular velocity of a two-pole machine. The electromagnetic torque for a P-pole machine is, therefore,

$$T_e = \frac{3}{2}\frac{P}{2} L_m \mathrm{Im}\left[\vec{i}_{qds} \vec{i}_{qdr}^*\right] = \frac{3}{2}\frac{P}{2} L_m \left(i_{qs}i_{dr} - i_{ds}i_{qr}\right) \tag{5.50}$$

Several alternative forms of electromagnetic torque can be derived using the stator and rotor flux linkage expressions. A couple of these torque expressions are as follows:

$$T_e = \frac{3}{2}\frac{P}{2} \mathrm{Im}\left[\vec{i}_{qds} \vec{\lambda}_{qds}^*\right] = \frac{3}{2}\frac{P}{2} \left(\lambda_{ds}i_{qs} - \lambda_{qs}i_{ds}\right) \tag{5.51}$$

$$T_e = \frac{3}{2}\frac{P}{2}\frac{L_m}{L_r} \mathrm{Im}\left[\vec{i}_{qds} \vec{\lambda}_{qdr}^*\right] = \frac{3}{2}\frac{P}{2}\frac{L_m}{L_r}\left(\lambda_{dr}i_{qs} - \lambda_{qr}i_{ds}\right) \tag{5.52}$$

The torque expressions in terms of dq variables are used in the vector control of induction motor drives. Vector control implementations are accomplished in one of the several available choices of reference frames, such as rotor flux-oriented reference frame, stator flux-oriented reference frame, or air gap flux-oriented reference frame. The abc variables at the input of the controller are converted to dq variables in the chosen reference frame at the input of the controller. The control computations take place in terms of dq variables, and the generated command outputs are again converted back to abc variables. The inverter controller executes the commands to establish the desired currents or voltages in the drive system. Some of these drives will be discussed in Chapter 8.

Torque expression derived in this section is useful for motor controller implementation, while the simplified torque expression of Equation 5.34 is useful for system level analysis.

REFERENCES

1. Dubey, G., *Power Semiconductor Controlled Drives*, Prentice Hall, New York, 1989.
2. Park, R.H., Two-Reaction Theory of Synchronous Machines — Generalized Method of Analysis — Part I, *AIEE Transactions*, Vol. 48, July, New York, 1929, pp. 716–727.
3. Vas, P., *Electric Machines and Drives: A Space-Vector Theory Approach*, Oxford University Press, Oxford, 1992.
4. Novotny, D.W. and Lipo, T.A., *Vector Control and Dynamics of AC Drives*, Oxford University Press, Oxford, 1996.
5. Mohan, N., *Electric Drives — An Integrated Approach*, MNPERE, Minneapolis, MN, 2001.

6. Lyon, W.V., *Transient Analysis of Alternating Current Machinery*, John Wiley & Sons, New York, 1954.
7. Krasue, P.C. and Wasynchuk, O., *Analysis of Electric Machinery*, McGraw-Hill, New York, 1986.

PROBLEMS

5.1

Find the condition of operation that minimizes the losses in a separately excited DC machine. (Start by writing an equation for P_{loss} in terms of the field currents and armature currents. Assuming linearity for all the nonlinear functions, establish the relation between armature current and field current, and then find the condition for minimum P_{loss}.)

5.2

Present an argument why it is impossible to achieve maximum efficiency at every operating point (T^*, ω^*) for a permanent magnet DC machine. (Start by writing an equation for P_{loss} in terms of T, ω, and machine flux ϕ).

5.3

Proceeding as in Problem 5.2, explain why it is impossible to minimize losses at any operating point (T^*, ω^*) for a series DC motor.

6 PM and SR Machines

6.1 PERMANENT MAGNET MACHINES

Machines that use magnets to produce air-gap magnetic flux instead of field coils, as in DC commutator machines, or the magnetizing component of stator current, as in induction machines, are permanent magnet (PM) machines. This configuration eliminates rotor copper loss as well as the need for maintenance of the field exciting circuit. PM machines can be broadly classified into two categories:

- *Synchronous machines (PMSMs)*: These machines have a uniformly rotating stator field as in induction machines. Induced waveforms are sinusoidal, and hence, *dq* transformation and vector control are possible.
- *Trapezoidal or square-wave machines*: These are also known as brushless DC or electronically commutated machines. Induced voltages are trapezoidal in nature. The stator field is switched in discrete steps with square-wave pulses.

There are several advantages of using permanent magnets for providing excitation in AC machines. Permanent magnets provide a loss-free excitation in a compact way without complications of connections to the external stationary electric circuits. This is especially true for smaller machines, because there is always an excitation penalty associated with providing the rotor field through electrical circuits. Large synchronous machines use rotor conductors to provide the excitation, because the losses in the exciter circuit, referred to as the excitation penalty, are small, especially when compared to the high costs of magnets. For smaller machines, the mmf required is small, and the resistive effects often become comparable and dominating, resulting in lower efficiency. The smaller cross-sectional area of the windings for small power machines further deteriorates the resistive loss effect. Moreover, the cross-sectional area available for winding decreases as the motor size gets smaller. The loss-free excitation of PM in smaller machines with a compact area is a plus, with the only drawback being the high costs of the permanent magnets. Nevertheless, PM machines are a strong contender for EV and HEV drives, despite their larger size. The factors guiding the trend are excellent performance and high power density achievable from PM drives. Cost is not yet the prime consideration, when acceptance of the EV and the HEV by the people is still at a hesitation stage.

6.1.1. PERMANENT MAGNETS

Permanent magnets are a source of mmf much like a constant current source, with relative permeability μ_r just greater than air, i.e., $\mu_r \approx 1.05 - 1.07$. PM characteristics

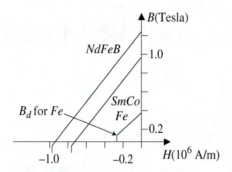

FIGURE 6.1 Characteristics of commonly used permanent magnets.

are displayed in the second quadrant of the *B-H* plot, as shown in Figure 6.1, confirming the fact that these are sources of mmf. The magnets remain permanent as long as the operating point is within the linear region of its *B-H* characteristics. However, if the flux density is reduced beyond the knee-point of the characteristics (B_d), some magnetism will be lost permanently. On removal of the demagnetizing field greater than the limit, new characteristics will include another straight line parallel to, but lower than, the original. The common type of magnets used in PM machines are the ferrites, samarium cobalt (SmCo), and neodymium-iron-boron (NdFeB). The features and properties of these three magnets are as follows.[1]

6.1.1.1   Ferrites

- They have been available for decades.
- Their cost is low.
- The residual flux density B_r at 0.3 to 0.4 T is much lower than the desired range of gap flux density for high power density.
- B_d is higher for those ferrites for which B_r is higher.
- Ferrites have high resistivity and low core losses.
- They can be operational up to 100°C.
- An increase in temperature increases B_r and decreases B_d.

6.1.1.2   Samarium Cobalt (SmCo)

- This material has a higher value of B_r, 0.8 to 1.1 T.
- B_d extends well into the third quadrant.
- B_r decreases somewhat with temperature, while B_d increases. This leads to increased sensitivity to demagnetization as temperature increases.
- Resistivity is 50 times that of Cu.
- Cost is relatively high, reflecting the cost of a rare earth element and an expensive metal.

6.1.1.3 Neodymium-Iron-Boron (NdFeB)

- Sintered NdFeB was developed in Japan in 1983 and provided the major impetus to PM motors.
- B_r is in the range of 1.1 to 1.25 T at room temperature. This is adequate to produce a flux density of 0.8 to 0.9 T across a relatively large air gap.
- B_r decreases by about 0.1% for each degree rise in temperature.
- The knee-point of flux density (B_d) increases rapidly with temperature. This imposes a limit on maximum temperature for NdFeB in the range of 100 to 180°C, depending on the detailed composition.
- The cost of these sintered NdFeB materials is still high, mainly because of the manufacturing complexity of the sintering process.
- The cost may be reduced in the future with an increase in volume use. Fe (77%) and B (8%) cost relatively little, and Nd is one of the more prevalent rare earth elements.
- Bonded NdFeB magnets can be produced at a lower cost, but B_r is lower, at about 0.6 to 0.7 T.

PM machines are designed with adequate considerations for magnet protection. Demagnetization may occur if flux density is reduced below the knee-point of flux density B_d. Most PM motors are designed to withstand considerable overload currents (two to four times the rated) without danger to the magnets.

6.1.2 PM SYNCHRONOUS MOTORS

The permanent magnet synchronous motor (PMSM), also known as the sinusoidal brushless DC motor or PM DC motor is a synchronous motor, where the field mmf is provided with permanent magnets. The PMSM has high efficiency and a cooling system that is easier to design. The use of rare earth magnet materials increases the flux density in the air gap and, accordingly, increases the motor power density and torque-to-inertia ratio. In high-performance motion control systems that require servo-type operation, the PMSM can provide fast response, high power density, and high efficiency. In certain applications like robotics and aerospace actuators, it is preferable to have the weight as low as possible for a given output power. The PMSM, similar to induction and DC machines, is fed from a power electronic inverter for efficient operation of the system. Smooth torque output is maintained in these machines by shaping the motor currents, which necessitates a high-resolution position sensor and current sensors. The control algorithm is implemented in a digital processor using feedback from the sensors. A flux weakening operation that enables a constant power mode of operation is possible in PMSM by applying a stator flux in opposition to the rotor magnet flux. The motor high speed limit depends on the motor parameters, its current rating, the back-emf waveform, and the maximum output voltage of the inverter.

The permanent magnets in PMSM machines are not only expensive, but they are also sensitive to temperature and load conditions, which constitute the major drawbacks of PM machines. Most of the PMSMs are found in small to medium power applications, although there are some high-power applications for which PMSMs are used.

PMSM and induction motors have good performance in terms of torque response and have rugged motor structures, although broken magnet chips in PM machines is a concern. The slip speed calculation makes the induction motor control more complicated than that of the PMSM, as the latter has only stationary and synchronous reference frames. Without a rotor cage, PMSM has a lower inertia that helps the electrical response time, although the induction motor electrical response characteristics will be the fastest because of the smaller time constant. The electrical time constant of magnetic circuits is determined by the L/R ratio. The load current transient in induction machines is limited only by small leakage inductance, where the time constant inductance in PM machines is the much higher self-inductance. With a higher power density, the PMSM is smaller in size compared to an induction motor with the same power rating. PMSM is more efficient and easier to cool, compared to the induction machines, due to the absence of rotor copper loss. The induction motor has lower cost and cogging torque because of the absence of permanent magnets.

Rotor temperature variation is a big problem for the induction motor because it causes detuning of the field-oriented controller. The induction motor can sustain high peak current at several times the rated current without the danger of demagnetizing the magnets. However, the PMSM can also safely sustain two to three times the rated current. The major difficulty with the PMSM occurs at high speed when the eddy current in the magnet (a sintered magnet has good conductivity) causes magnet heating and may cause demagnetization. To prevent such occurrences, the magnets are sometimes split into several pieces along the axial length, which increases resistance to eddy current.

The surface-mount PM machine suffers from poor field weakening capability. The efficiency of the machine is also low at high speed due to higher core and copper losses. However, the interior PM machine has excellent field weakening characteristics. The field weakening performance of the induction machine is also satisfactory, the range being two to three times the base speed.

6.1.2.1 Types of PMSMs

The PMSM has a stator with a set of three-phase sinusoidally distributed copper windings similar to the windings described in Section 5.4 on AC machines. A balanced set of applied three-phase voltages forces a balanced set of sinusoidal currents in the three-phase stator windings, which in turn establishes the constant amplitude rotating mmf in the air gap. Stator currents are regulated using rotor position feedback so that the applied current frequency is always in synchronism with the rotor. Permanent magnets in the rotor are appropriately shaped, and their magnetization directions are controlled such that the rotor flux linkage created is sinusoidal. Electromagnetic torque is produced at the shaft by the interaction of these two stator and rotor magnetic fields.

PMSMs are classified according to the position and shape of the permanent magnets in the rotors. Three common arrangements of the rotors are surface mounted,

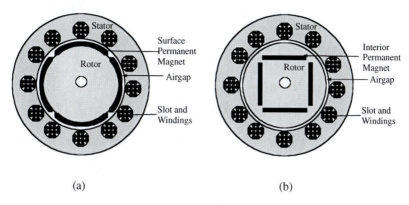

(a)   (b)

FIGURE 6.2 Permanent magnet machines: (a) surface-mounted; (b) interior.

inset, and interior or buried. The surface-mounted and interior PM machine configurations are shown in Figure 6.2. The difference between surface mounted and inset
magnets is that the magnets in the latter are inside the rotor surface but still exposed
to the air gap. The surface-mount and inset rotor PMSMs are often collectively called
the surface-mount PMSMs. The other type of PMSM is the interior PMSM, so
named because of the magnet arrangement in the rotor. In the surface-mounted
PMSM, the magnets are epoxy glued or wedge fixed to the cylindrical rotor. Non-
magnetic stainless steel or carbon fiber sleeves are also used to contain the magnets.
The manufacturing of this kind of rotor is simple, although the mechanical strength
of the rotor is only as good as that of the epoxy glue. The direct and quadrature-
axes inductances of the surface-mounted PMSM are approximately equal, because
permeability of the path that the flux crosses between the stator and the rotor is
equal around the stator circumference. Uniformity in the magnetic path despite the
presence of magnets is because the permeability of magnets is approximately equal
to that of the air. The space needed to mount the magnets increases the radial distance
of the effective air gap, making self-inductance relatively smaller in PMSMs. In the
inset PMSM, magnets are put into the rotor surface slots, which makes them more
secured. The direct and quadrature axes reluctances are unequal in inset PMSMs,
because space is occupied by a magnet in the direct axis and by iron in the quadrature
axis. The quadrature axis inductance L_q is larger than the direct axis inductance L_d,
because the direct axis flux path has a larger effective air gap and, hence, higher
reluctance, although the length of the air gap between the stator and the rotor is the
same. The interior PMSM has its magnets buried inside the rotor. The manufacturing
process is complicated and expensive for the interior PMSM. The quadrature axis
inductance L_q in interior PMSMs can be much larger than the direct axis inductance
L_d. The larger difference in the d- and q-axes inductances make the interior PM
more suitable for flux weakening operation, delivering a wider constant power region
compared to the other types of PMSMs. The extended constant power range capability is extremely important for EV and HEV applications to eliminate the use of
multiple gear ratios and to reduce the power inverter volt-ampere rating. Because
of the unequal reluctance paths in the direct and quadrature axes, a reluctance torque
exists in buried and inset PMSMs.

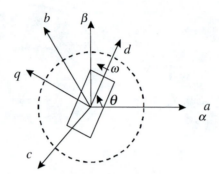

FIGURE 6.3 The stationary and synchronous frames in PMSM.

6.1.3 PMSM Models

The modeling of a PMSM can be accomplished in the stationary reference frame or in the synchronous reference frame, as shown in Figure 6.3. The *abc* and αβ reference frames are fixed in the stator with the direction of α-axes chosen in the same direction as *a*-axes. The β-axis lags the α-axis by 90° of space angle. The *abc* to αβ transformation essentially transforms the three-phase stationary variables to a set of two-phase stationary variables. The *dq* reference frame is locked to the rotor frame. The *d*-axis is aligned with the magnet flux direction, while the *q*-axis lags the *d*-axis by 90° of space angle.

6.1.3.1 Voltage Equations

The stator circuit of a PMSM is similar to that of an induction motor or a wound rotor synchronous motor, with the applied voltage being balanced by the stator winding resistance drop and the induced voltage in the winding. The PMSM model will be derived in the following, assuming that the stator windings in the three phases are symmetrical and sinusoidally distributed. Eddy current and hysteresis losses will be neglected in the model, and damper or cage windings will not be considered. Damper windings are not necessary in PMSMs, because these machines are always soft-started by lowering the inverter frequency. Large armature currents can be tolerated in these machines without significant demagnetization. The stator phase voltage equation in the stationary *abc* reference frame is

$$\vec{v}_{abcs} = \overline{R}_s \vec{i}_{abcs} + \frac{d}{dt} \vec{\lambda}_{abcs}$$

where

$$\left(\vec{f}_{abcs}\right)^T = \begin{bmatrix} f_{as} & f_{bs} & f_{cs} \end{bmatrix}$$

and

$$\overline{R}_s = \text{diag}\left[R_s \ R_s \ R_s\right]$$

The flux linkages are

$$\overline{\lambda}_{abcs} = \overline{L}_s \overline{i}_{abcs} + \overline{\lambda}_f$$

$\overline{L}_s$ is the inductance matrix that is the same for any synchronous machine. $\overline{\lambda}_f$ is due to the permanent magnet and is given by

$$\overline{\lambda}_f = \lambda_f \begin{bmatrix} \sin\theta \\ \sin\left(\theta_r - \dfrac{2\pi}{3}\right) \\ \sin\left(\theta_r + \dfrac{2\pi}{3}\right) \end{bmatrix}$$

λ_f is the amplitude of the flux linkage established by the permanent magnet, as viewed from the stator phase windings.

6.1.3.2 Voltage and Torque in Reference Frames

The *abc* variables can be transformed into the rotor reference frame, which is also the synchronous reference frame, for synchronous machines in steady state. The stator *dq* equations of the PMSM in the *dq* or rotor reference frame are as follows:

$$v_q = R_s i_q + \frac{d}{dt}\lambda_q + \omega_r \lambda_d$$

$$v_d = R_s i_d + \frac{d}{dt}\lambda_d - \omega_r \lambda_q$$

(6.1)

where

$$\lambda_q = L_q i_q$$
$$\lambda_d = L_d i_d + \lambda_f$$

Here, i_d and i_q are the *dq* axis stator currents, v_d and v_q are the *dq* axis stator voltages, R_s is the stator phase resistance, L_d and L_q are the *dq* axis phase inductances, λ_d and λ_q are *dq* axis flux linkages, and ω_r is the rotor speed in electrical rad/s. The subscript *s* used in Chapter 5 to refer to stator quantities for induction machines has been dropped here for simplicity. The d and q axis inductances are $L_d = L_{ls} + L_{md}$ and $L_q = L_{ls} + L_{mq}$. (L_{ls} is the leakage inductance.) Note that the d- and q-axes mutual inductances can be different in the case of PM machines. The electromagnetic torque is

$$T_e = \frac{3}{2}\frac{P}{2}\left[\lambda_f i_q + \left(L_d - L_q\right)i_d i_q\right]$$ (6.2)

where P is the number of poles.

Rotor position information gives the position of d- and q-axes. The control objective is to regulate the voltages v_d and v_q or the currents i_d and i_q by controlling the firing angles of the inverter switches. The rotor position is given by

$$\theta_r = \int_0^t \omega_r(\xi)d\xi + \theta_r(0)$$

6.1.3.3 *dq* and *αβ* Model

The state-space representation of the PMSM model is useful for the computer simulation of the motor. Representation in the commonly used *dq* synchronous reference frame is as follows:

$$\frac{di_d}{dt} = -\frac{R_s}{L_d}i_d + \frac{L_q}{L_d}\frac{P}{2}\omega i_q + \frac{v_d}{L_d}$$

$$\frac{di_q}{dt} = -\frac{R_s}{L_q}i_q - \frac{L_d}{L_q}\omega i_d - \frac{1}{L_q}\omega\lambda_f + \frac{v_q}{L_q}$$

$$\frac{d\omega}{dt} = \frac{1}{J}\left(T_e - T_l - B\omega - F\frac{\omega}{|\omega|}\right)$$ (6.3)

$$\frac{d\theta}{dt} = \frac{P}{2}\omega_r$$

where J is the rotor and load inertia, F is the coulomb friction, B is the viscous load, T_l is the load torque, P is the number of poles, ω_r is the mechanical rotor speed, and θ_r is the rotor position. The electrical rotor speed ω is related to the mechanical rotor speed as

$$\omega = \frac{P}{2}\omega_r$$

The stationary *αβ* reference frame is also sometimes used for modeling the PMSM. The *αβ* model is

$$\frac{di_\alpha}{dt} = -\frac{R_s}{L} i_\alpha + \frac{k_e}{L} \frac{P}{2} \omega_r \sin(\theta) + \frac{v_\alpha}{L}$$

$$\frac{di_\beta}{dt} = \frac{R_s}{L} i_\beta - \frac{k_e}{L} \frac{P}{2} \omega_r \cos(\theta) + \frac{v_\beta}{L}$$

$$\frac{d\omega}{dt} = -\frac{3}{2} \frac{k_e}{J} i_\alpha \sin(\theta) + \frac{3}{2} \frac{k_e}{J} i_\beta \cos(\theta) - \frac{T_l}{J} - \frac{B}{J}\omega - \frac{F}{J} \frac{\omega}{|\omega|}$$

$$\frac{d\theta}{dt} = \frac{P}{2} \omega_r$$

(6.4)

where v_α and v_β are the $\alpha\beta$-stator voltages, i_α and i_β are the $\alpha\beta$-stator currents, L is the phase inductance with $L \approx 0.5*(L_d + L_q)$ and k_e is the motor torque constant.

6.1.3.4 Transformation Equations

The transformation between the reference frames are given by Park's equations as

$$\begin{bmatrix} i_d \\ i_q \\ 0 \end{bmatrix} = T_{abc->dq} \begin{bmatrix} i_a \\ i_b \\ i_c \end{bmatrix}, \begin{bmatrix} i_\alpha \\ i_\beta \\ 0 \end{bmatrix} = T_{abc->\alpha\beta} \begin{bmatrix} i_a \\ i_b \\ i_c \end{bmatrix}, \begin{bmatrix} i_d \\ i_q \end{bmatrix} = T_{\alpha\beta->dq} \begin{bmatrix} i_\alpha \\ i_\beta \end{bmatrix}$$

(6.5)

where

$$T_{abc->dq} = \frac{2}{3} \begin{bmatrix} \cos(\theta) & \cos(\theta - 2\pi) & \cos(\theta + 2\pi/3) \\ \sin(\theta) & \sin(\theta - 2\pi/3) & \sin(\theta + 2\pi/3) \\ 0.5 & 0.5 & 0.5 \end{bmatrix}$$

$$T_{abc->\alpha\beta} = \frac{2}{3} \begin{bmatrix} 1 & -0.5 & -0.5 \\ 0 & -\sqrt{3}/2 & \sqrt{3}/2 \\ 0.5 & 0.5 & 0.5 \end{bmatrix}, T_{\alpha\beta->dq} = \begin{bmatrix} \cos(\theta) & \sin(\theta) \\ -\sin(\theta) & \cos(\theta) \end{bmatrix}$$

(6.6)

and

$$T_{dq->abc} = T^{-1}_{abc->dq}, T_{\alpha\beta->abc} = T^{-1}_{abc->\alpha\beta}, T_{dq->\alpha\beta} = T^{-1}_{\alpha\beta->dq}$$

The *abc* variables are obtained from the *dq* variables through the inverse of the Park transformation equations:

$$T_{dq->abc} = \begin{bmatrix} \cos(\theta) & \sin(\theta) & 1 \\ \cos(\theta - 2\pi/3) & \sin(\theta - 2\pi/3) & 1 \\ \cos(\theta + 2\pi/3) & \sin(\theta + 2\pi/3) & 1 \end{bmatrix}$$

(6.7)

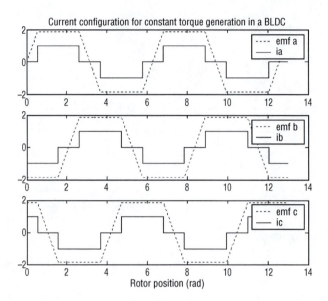

FIGURE 6.4 Back-emf and ideal phase currents in the three phases of a PM brushless DC motor.

6.1.4 PM BRUSHLESS DC MOTORS

Permanent magnet AC machines with trapezoidal back-emf waveforms are known as the PM brushless DC machines. The concentrated windings of the machine instead of the sinusoidally distributed windings on the stator are the reason for the trapezoidal-shaped back-emf waveforms. PM brushless DC motors are widely used in a range of applications, from computer drives to sophisticated medical equipment. The reason behind the popularity of these machines is the simplicity of control. Only six discrete rotor positions per electrical revolution are needed in a three-phase machine to synchronize the phase currents with the phase back-emfs for effective torque production. A set of three Hall sensors mounted on the stator facing a magnet wheel fixed to the rotor and placed 120° apart can easily give this position information. This eliminates the need for the high-resolution encoder or position sensor required in PM synchronous machines, but the penalty paid for position sensor simplification is in performance. For EV/HEV applications, a high-resolution encoder/resolver may be necessary for phase advancing at high speeds. Vector control is not possible in PM brushless DC machines because of the trapezoidal shape of back-emfs.

The three-phase back-emf waveforms and the ideal phase currents of a PM brushless DC motor are shown in Figure 6.4. The back-emf waveforms are fixed with respect to rotor position. Square-wave phase currents are supplied such that they are synchronized with the back-emf peak of the respective phase. The controller achieves this objective using rotor position feedback information. The motor basically operates like a DC motor, with such a controller configuration, from a control point of view; hence, the motor is designated as a brushless DC motor.

6.1.4.1 Brushless DC Motor Modeling

The permanent magnet in the rotor can be regarded as a constant current source, giving rise to the back-emfs in the stator windings. The three stator windings for the three phases are assumed to be identical, with 120° (electrical) phase displacement among them. Therefore, the stator winding resistances and the self-inductance of each of the three phases can be assumed to be identical. Let R_s be the stator phase winding resistance, $L_{aa} = L_{bb} = L_{cc} = L$ be the stator phase self-inductance, and $L_{ab} = L_{ac} = L_{bc} = M$ stand for stator mutual inductance.

The voltage balance equation for the three phases are as follows:

$$\begin{bmatrix} v_a \\ v_b \\ v_c \end{bmatrix} = R_s \cdot \begin{bmatrix} i_a \\ i_b \\ i_c \end{bmatrix} + \begin{bmatrix} L & M & M \\ M & L & M \\ M & M & L \end{bmatrix} \cdot p \cdot \begin{bmatrix} i_a \\ i_b \\ i_c \end{bmatrix} + \begin{bmatrix} e_a \\ e_b \\ e_c \end{bmatrix} \qquad (6.8)$$

where p is the operator d/dt, and e_a, e_b, and e_c are the back-emfs in the three phases. The back-emf is related to the phase flux-linkage as

$$e = \frac{d\lambda}{dt} = \frac{d\lambda}{d\theta_r} \cdot \frac{d\theta_r}{dt}$$

But, $d\theta_r/dt = \omega_r$, which is the speed of the rotor. Then,

$$e = \omega_r \cdot \frac{d\lambda}{d\theta_r} \qquad (6.9)$$

The back-emf will not have a trapezoidal shape during transient conditions. Similar to the back-emfs, the currents are also shifted by 120°, and they satisfy the condition $i_a + i_b + i_c = 0$. Therefore, we have $M \cdot i_b + M \cdot i_c = -M \cdot i_a$. Similar expressions exist for the two other phases. Equation (6.8) can then be simplified to

$$\begin{bmatrix} v_a \\ v_b \\ v_c \end{bmatrix} = R_s \cdot \begin{bmatrix} i_a \\ i_b \\ i_c \end{bmatrix} + \begin{bmatrix} L-M & 0 & 0 \\ 0 & L-M & 0 \\ 0 & 0 & L-M \end{bmatrix} \cdot p \cdot \begin{bmatrix} i_a \\ i_b \\ i_c \end{bmatrix} + \begin{bmatrix} e_a \\ e_b \\ e_c \end{bmatrix}$$

The rate of change of currents for applied voltages can be expressed as

$$p \cdot \begin{bmatrix} i_a \\ i_b \\ i_c \end{bmatrix} = \frac{1}{L-M} \cdot \left(\begin{bmatrix} v_a \\ v_b \\ v_c \end{bmatrix} - R_s \cdot \begin{bmatrix} i_a \\ i_b \\ i_c \end{bmatrix} - \begin{bmatrix} e_a \\ e_b \\ e_c \end{bmatrix} \right) \qquad (6.10)$$

The electrical power transferred to the rotor is equal to the mechanical power $T_e \omega_r$ available at the shaft. Using this equality, the electromagnetic torque for the PM brushless DC motor is

$$T_e = \frac{e_a \cdot i_a + e_b \cdot i_b + e_c \cdot i_c}{\omega_r} \qquad (6.11)$$

For the strategy described previously, where only two phase currents were active at one time, the torque expression for equal currents in two phases simplifies to

$$T_e = \frac{2 \cdot e_{max} \cdot I}{\omega_r} \qquad (6.12)$$

Because the currents are controlled to synchronize with the maximum back-emf only, e_{max} has been used in Equation 6.12 instead of e as a function of time or rotor position. Assuming magnetic linearity, Equation 6.9 can be written as

$$e = K \cdot \omega_r \cdot \frac{dL}{d\theta}$$

Hence, the maximum back-emf is

$$e_{max} = K \cdot \left[\frac{dL}{d\theta}\right]_{max} \cdot \omega_r$$

or

$$e_{max} = K' \cdot \omega_r \qquad (6.13)$$

Equations 6.12 and 6.13 are similar to the $E = K \cdot \phi \cdot \omega$ and $T = K \cdot \phi \cdot I$ equations associated with regular DC machines. Therefore, with the described control strategy, PM brushless DC motors can be considered as behaving like a DC machine.

6.2 SWITCHED RELUCTANCE MACHINES

6.2.1 SRM CONFIGURATION

The switched reluctance motor (SRM) is a doubly salient, singly excited reluctance machine with independent phase windings on the stator. The stator and the rotor are made of magnetic steel laminations, with the latter having no windings or magnets. Cross-sectional diagrams of a four-phase, 8/6 SRM and a three-phase, 12/8 SRM are shown in Figure 6.5. The three-phase, 12/8 machine is a two-repetition version

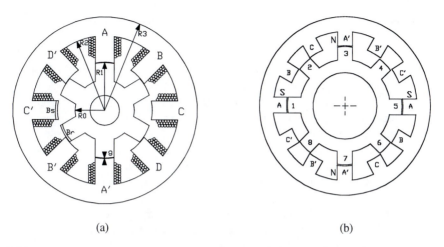

(a)   (b)

FIGURE 6.5 Cross-sections of three-phase SR machines: (a) four-phase 8/6 structure; (b) 12/8, two-repetition (two-channel) structure. (From Husain, I., Switched reluctance machines, in *The Power Electronics Handbook*, Skvarenina, T.L., Ed., CRC Press, Boca Raton, FL, 2002.)

of the basic 6/4 structure within the single stator geometry. The two-repetition machine can alternately be labeled as a four-poles/phase machine, compared to the 6/4 structure with two poles/phase. The stator windings on diametrically opposite poles are connected in series or in parallel to form one phase of the motor. When a stator phase is energized, the most adjacent rotor pole-pair is attracted toward the energized stator in order to minimize the reluctance of the magnetic path. Therefore, it is possible to develop constant torque in either direction of rotation by energizing consecutive phases in succession.

The aligned position of a phase is defined to be the orientation when the stator and rotor poles of the phase are perfectly aligned, attaining the minimum reluctance position. The unsaturated phase inductance is maximum (L_a) in this position. Phase inductance decreases gradually as the rotor poles move away from the aligned position in either direction. When the rotor poles are symmetrically misaligned with the stator poles of a phase, the position is said to be the unaligned position. The phase has the minimum inductance (L_u) in this position. Although the concept of inductance is not valid for a highly saturating machine like SRM, the unsaturated aligned and unaligned inductances are two key reference positions for the controller.

Several other combinations of the number of stator and rotor poles exist, such as 10/4, 12/8, etc. A 4/2 or a 2/2 configuration is also possible, but if the stator and rotor poles are aligned exactly, then it would be impossible to develop a starting torque. Configurations with a higher number of stator/rotor pole combinations have less torque ripple and do not have the problem of starting torque.

6.2.1.1 Advantages and Drawbacks

Switched reluctance machines or motors (SRM) possess unique features that make them strong competitors to existing AC and DC motors in various adjustable speed drives and servo applications. The advantages of an SRM can be summarized as follows:

- Simple and low-cost machine construction due to the absence of rotor winding and permanent magnets.
- No shoot-through faults between the DC buses in the SRM drive converter, because each phase winding is connected in series with converter switching elements.
- Bidirectional currents are not necessary, which facilitates the reduction of the number of power switches in certain applications.
- The bulk of the losses appears in the stator, which is relatively easier to cool.
- The torque–speed characteristics of the motor can be tailored to the application requirement more easily during the design stage than in the case of induction and PM machines.
- The starting torque can be very high without the problem of excessive in-rush current due to its higher self-inductance.
- The maximum permissible rotor temperature is higher, because there is no permanent magnet.
- They have low rotor inertia and high torque/inertia ratio.
- They make extremely high speeds with a wide constant power region possible.
- Independent stator phases enable drive operation in spite of the loss of one or more phases.

The SRM also comes with a few disadvantages, among which torque ripple and acoustic noise are the most critical. The double saliency construction and the discrete nature of torque production by the independent phases lead to higher torque ripple compared to other machines. The higher torque ripple also causes the ripple current in the DC supply to be quite large, necessitating a large filter capacitor. The doubly salient structure of the SRM also causes higher acoustic noise compared to other machines. The main source of acoustic noise is the radial magnetic force-induced resonance with the circumferential mode shapes of the stator. Among other disadvantages of SR motors are special converter and higher terminal connection requirements that add cost to the system.

The absence of permanent magnets imposes the burden of excitation on the stator windings and converter, which increases the converter kVA requirement. Compared to PM brushless machines, the per-unit stator copper losses will be higher, reducing the efficiency and torque per ampere. However, maximum speed at constant power is not limited by the fixed magnet flux as in the PM machine, and hence, an extended constant power region of operation is possible in SRMs. The control can be simpler than the field-oriented control of induction machines, although for torque ripple minimization, significant computations may be required for an SRM drive.

6.2.2 Basic Principle of Operation

6.2.2.1 Voltage-Balance Equation

The general equation governing the flow of stator current in one phase of an SRM can be written as

$$V_{ph} = i_{ph}R_s + \frac{d\lambda_{ph}}{dt} \qquad (6.14)$$

where V_{ph} is the DC bus voltage, i_{ph} is the instantaneous phase current, R_s is the winding resistance, and λ_{ph} is the flux linking the coil. The SRM is always driven into saturation to maximize the utilization of the magnetic circuit and, hence, the flux-linkage λ_{ph} is a nonlinear function of stator current and rotor position:

$$\lambda_{ph} = \lambda_{ph}\left(i_{ph}, \theta\right)$$

The electromagnetic profile of an SRM is defined by the $\lambda - i - \theta$ characteristics shown in Figure 6.6. The stator phase voltage can be expressed as

$$V_{ph} = i_{ph}R_s + \frac{\partial\lambda_{ph}}{\partial i_{ph}}\frac{di_{ph}}{dt} + \frac{\partial\lambda_{ph}}{\partial\theta}\frac{d\theta}{dt} = i_{ph}R_s + L_{inc}\frac{di_{ph}}{dt} + k_v\omega \qquad (6.15)$$

where L_{inc} is the incremental inductance, k_v is the current dependent back-emf coefficient, and $\omega = d\theta/dt$ is the rotor angular speed. Assuming magnetic linearity [where $\lambda_{ph} = L_{ph}(\theta)i_{ph}$], the voltage expression can be simplified as

$$V_{ph} = i_{ph}R_s + L_{ph}(\theta)\frac{di_{ph}}{dt} + i_{ph}\frac{dL_{ph}(\theta)}{dt}\omega \qquad (6.16)$$

The last term in Equation (6.16) is the "back-emf" or "motional-emf," and it has the same effect on SRM as the back-emf has on DC motors or electronically commutated motors (ECMs). However, the back-emf in SRM is generated in a different way from the DC machines or ECMs, where it is caused by a rotating magnetic field. In an SRM, there is no rotor field, and back-emf depends on the instantaneous rate of change of phase flux-linkage.

In the linear case, which is always valid for lower levels of phase current, the per-phase equivalent circuit of an SRM consists of a resistance, an inductance, and a back-emf component. The back-emf vanishes when there is no phase current or the phase inductance is constant relative to the rotor position. Depending on the magnitude of current and rotor angular position, the equivalent circuit changes its structure from being primarily an R-L circuit to primarily a back-emf dependent circuit.

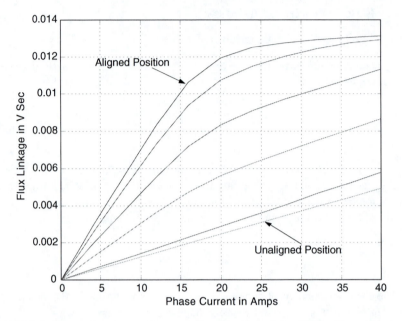

FIGURE 6.6 Flux-angle-current characteristics of a four-phase SRM. (From Husain, I., Switched reluctance machines, in *The Power Electronics Handbook*, Skvarenina, T.L., Ed., CRC Press, Boca Raton, FL, 2002.)

6.2.2.2 Energy Conversion

The energy conversion process in an SRM can be evaluated using the power balance relationship. Multiplying Equation 6.16 by i_{ph} on both sides, the instantaneous input power can be expressed as

$$P_{in} = V_{ph}i_{ph} = i_{ph}^2 R_s + \left(L_{ph}i_{ph}\frac{di_{ph}}{dt} + \frac{1}{2}i_{ph}^2\frac{dL_{ph}}{d\theta}\omega \right) + \frac{1}{2}i_{ph}^2\frac{dL_{ph}}{d\theta}\omega$$

$$= i_{ph}^2 R + \frac{d}{dt}\left(\frac{1}{2}L_{ph}i_{ph}^2 \right) + \frac{1}{2}i_{ph}^2\frac{dL_{ph}}{d\theta}\omega$$

(6.17)

The first term represents the stator winding loss; the second term denotes the rate of change of magnetic stored energy; and the third term is the mechanical output power. The rate of change of magnetic stored energy always exceeds the electromechanical energy conversion term. The most effective use of the energy supplied is when the current is maintained constant during the positive $dL_{ph}/d\theta$ slope. The magnetic stored energy is not necessarily lost; it can be retrieved by the electrical source if an appropriate converter topology is used. In the case of a linear SRM, energy conversion effectiveness can be, at most, 50%, as shown in the energy division diagram of Figure 6.7a. The drawback of lower effectiveness is the increase in converter volt-amp rating for a given power conversion of the SRM. The division

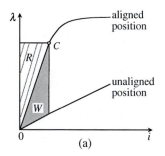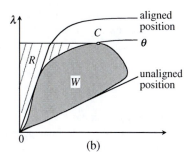

FIGURE 6.7 Energy partitioning during one complete working stroke: (a) linear assumption; (b) practical case. W = energy converted into mechanical work. R = energy returned to the DC supply. (From Husain, I., Switched reluctance machines, in *The Power Electronics Handbook*, Skvarenina, T.L., Ed., CRC Press, Boca Raton, FL, 2002.)

of input energy increases in favor of energy conversion if the motor operates under magnetic saturation. The energy division under saturation is shown in Figure 6.7b. This is the primary reason for always operating the SRM under saturation. The term "energy ratio" instead of "efficiency" is often used for SRM because of the unique situation of the energy conversion process. The energy ratio is defined as

$$ER = \frac{W}{W + R} \tag{6.18}$$

where W is the energy converted into mechanical work, and R is the energy returned to the source using a regenerative converter. The term "energy ratio" is analogous to the term "power factor" used for AC machines.

6.2.2.3 Torque Production

Torque is produced in the SRM by the tendency of the rotor to attain the minimum reluctance position when a stator phase is excited. The general expression for instantaneous torque for such a device that operates under the reluctance principle is as follows:

$$T_{ph}\left(\theta, i_{ph}\right) = \left.\frac{\partial W'\left(\theta, i_{ph}\right)}{\partial \theta}\right|_{i = \text{constant}} \tag{6.19}$$

where W' is the co-energy defined as

$$W' = \int_{0}^{i} \lambda_{ph}\left(\theta, i_{ph}\right) di$$

Obviously, instantaneous torque is not constant. Total instantaneous torque of the machine is given by the sum of the individual phase torques:

$$T_{inst}(\theta,i) = \sum_{phases} T_{ph}(\theta, i_{ph})$$

(6.20)

SRM electromechanical properties are defined by the static $T-i-\theta$ characteristics of a phase, an example of which is shown in Figure 6.8. Average torque is a more important parameter from the user's perspective and can be derived mathematically by integrating Equation 6.20:

$$T_{avg} = \frac{1}{T} \int_{0}^{T} T_{inst} dt$$

(6.21)

Average torque is also an important parameter during the design process.

When magnetic saturation can be neglected, instantaneous torque expression becomes:

$$T_{ph}(\theta,i) = \frac{1}{2} i_{ph}^2 \frac{dL_{ph}(\theta)}{d\theta}$$

(6.22)

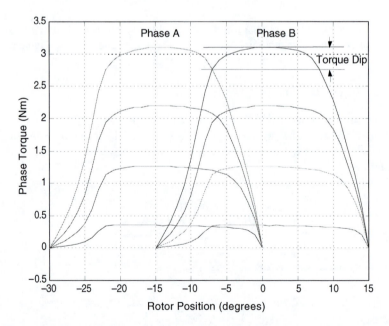

FIGURE 6.8 Torque-angle-current characteristics of a four-phase SRM for four constant current levels. (From Husain, I., Switched reluctance machines, in *The Power Electronics Handbook*, Skvarenina, T.L., Ed., CRC Press, Boca Raton, FL, 2002.)

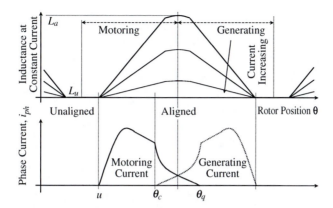

FIGURE 6.9 Phase currents for motoring and generating modes with respect to rotor position and idealized inductance profiles. (From Husain, I., Switched reluctance machines, in *The Power Electronics Handbook*, Skvarenina, T.L., Ed., CRC Press, Boca Raton, FL, 2002.)

The linear torque expression also follows from the energy conversion term (last term) in Equation 6.17. The phase current needs to be synchronized with the rotor position for effective torque production. For positive or motoring torque, the phase current is switched so that the rotor is moving from the unaligned position toward the aligned position. The linear SRM model is insightful in understanding these situations. Equation 6.22 clearly shows that for motoring torque, the phase current must coincide with the rising inductance region. On the other hand, the phase current must coincide with the decreasing inductance region for braking or generating torque. The phase currents for motoring and generating modes of operation are shown in Figure 6.9, with respect to the phase inductance profiles. Torque expression also shows that the direction of current is immaterial in torque production. The optimum performance of the drive system depends on the appropriate positioning of phase currents relative to the rotor angular position. Therefore, a rotor position transducer is essential to provide the position feedback signal to the controller.

6.2.2.4 Torque-Speed Characteristics

The torque–speed plane of an SRM drive can be divided into three regions, as shown in Figure 6.10. The constant torque region is the region below the base speed ω_b, which is defined as the highest speed when maximum rated current can be applied to the motor at rated voltage with fixed firing angles. In other words, ω_b is the lowest possible speed for the motor to operate at its rated power.

   Region 1: In the low speed region of operation, the current rises almost instantaneously after turn on, because the back-emf is small. The current can be set at any desired level by means of regulators, such as hysteresis controller or voltage pulse width modulation (PWM) controller.

   As the motor speed increases, the back-EMF soon becomes comparable to the DC bus voltage, and it is necessary to phase advance the turn-on angle so that the current can rise to the desired level against a lower

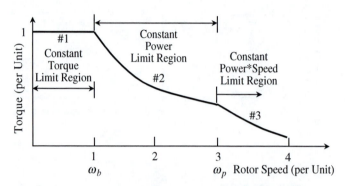

FIGURE 6.10 Torque–speed characteristics of an SRM drive. (From Husain, I., Switched reluctance machines, in *The Power Electronics Handbook*, Skvarenina, T.L., Ed., CRC Press, Boca Raton, FL, 2002.)

back-EMF. Maximum current can still be forced into the motor by PWM or chopping control to maintain the maximum torque production. The phase excitation pulses also need to be turned off a certain time before the rotor passes alignment to allow the freewheeling current to decay so that no braking torque is produced.

Region 2: When the back-EMF exceeds the DC bus voltage in high-speed operation, the current starts to decrease once pole overlap begins and PWM or chopping control is no longer possible. The natural characteristic of the SRM, when operated with fixed supply voltage and fixed conduction angle θ_{dwell} (also known as the dwell angle), begins when the phase excitation time falls off inversely with speed and so does the current. Because torque is roughly proportional to the square of the current, the rapid fall in torque with speed is countered by adjusting the conduction angle. Increasing the conduction angle increases the effective amps delivered to the phase. Torque production is maintained at a level high enough in this region by adjusting the conduction angle θ_{dwell} with the single-pulse mode of operation. The controller maintains the torque inversely proportional to the speed; hence, this region is called the constant power region. The conduction angle is also increased by advancing the turn-on angle until the θ_{dwell} reaches its upper limit at speed ω_p.

The medium speed range through which constant power operation can be maintained is quite wide, and high maximum speeds can be achieved.

Region 3: The θ_{dwell} upper limit is reached when it occupies half the rotor pole-pitch, i.e., half the electrical cycle. θ_{dwell} cannot be increased further, because the flux would not return to zero, and the current conduction would become continuous. The torque in this region is governed by natural characteristics, falling off as $1/\omega^2$.

The torque–speed characteristics of the SRM are similar to a DC series motor, which is not surprising, considering that the back-emf is proportional to current, while the torque is proportional to the square of the current.

6.2.3   SRM Design

6.2.3.1   Basic Design Guidelines

The fundamental design rules governing the choice of phase numbers, pole numbers, and pole arcs are discussed in detail by Lawrenson et al.[2] and also by Miller.[3] From a designer's point of view, the objectives are to minimize the core losses, to have good starting capability, to minimize the unwanted effects due to varying flux distributions and saturation, and to eliminate mutual coupling. The choice of the number of phases and poles is open, but a number of factors need to be evaluated prior to making a selection. A comprehensive design methodology of SRM appears in the literature.[4]

The fundamental switching frequency is given by

$$f = \frac{n}{60} N_r \ \text{Hz} \tag{6.23}$$

where n is the motor speed in rev/m, and N_r is the number of rotor poles. The "step angle" or "stroke" of an SRM is given by

$$\text{Step angle } \varepsilon = \frac{2\pi}{N_{ph}.N_{rep}.N_r} \tag{6.24}$$

The stroke angle is an important design parameter related to the frequency of control per rotor revolution. N_{rep} represents the multiplicity of the basic SRM configuration, which can also be stated as the number of pole pairs per phase. N_{ph} is the number of phases. N_{ph} and N_{rep} together set the number of stator poles.

The regular choice of the number of rotor poles in an SRM is

$$N_r = N_s \pm k_m \tag{6.25}$$

where k_m is an integer such that $k_m \bmod q \neq 0$ and N_s is the number of stator poles. Some combinations of parameters allowed by Equation 6.25 are not feasible, because sufficient space must exist between the poles for the windings. The most common choice of Equation 6.25 for the selection of stator and rotor pole numbers is $k_m = 2$ with the negative sign.

The highest possible saliency ratio (the ratio between the maximum and minimum unsaturated inductance levels) is desired to achieve the highest possible torque per ampere, but as the rotor and stator pole arcs are decreased, the torque ripple tends to increase. The torque ripple adversely affects the dynamic performance of an SRM drive. For many applications, it is desirable to minimize the torque ripple, which can be partially achieved through appropriate design. The torque dip observed in the $T - i - \theta$ characteristics of an SRM (see Figure 6.8) is an indirect measure of the torque ripple expected in the drive system. The torque dip is the difference between the peak torque of a phase and the torque at an angle, where two overlapping phases produce equal torque at equal levels of current. The smaller the torque dip, the less the torque ripple. The $T - i - \theta$ characteristics of the SRM depend on the stator–rotor pole overlap angle, pole geometry, material properties, number of poles, and number of phases. A design trade-off needs to be considered to achieve the desired goals. The $T - i - \theta$ characteristics must be studied through finite element analysis during the design stage to evaluate the peak torque and torque dip values.

REFERENCES

1. Miller, T.J.E., *Brushless Permanent Magnet and Switched Reluctance Motor Drives*, Oxford University Press, Oxford, 1989.
2. Lawrenson, P.J. et al., Variable-speed switched reluctance motors, *IEE Proc.*, Pt. B, Vol. 127, No. 4, July, 1980, pp. 253–265.
3. Miller, T.J.E., *Switched Reluctance Motors and Their Control,* Magna Physics Publishing, Hillsboro, OH; Oxford Science Publications, Oxford, UK, 1993.
4. Anwar, M.N., Husain, I., and Radun, A.V., A Comprehensive Design Methodology for Switched Reluctance Machines, *IEEE-IAS Annual Conf.,* Italy, October, 2000.

PROBLEMS

6.1

(a) A PM brushless DC has a torque constant of 0.12 N-m/A referred to the DC supply. Estimate its no-load speed in rpm when connected to a 48 V DC supply.
(b) If the armature resistance is 0.15 Ω/phase and the total voltage drop in the controller transistors is 2 V, determine the stall current and the stall torque.

6.2

Consider a three-phase 6/8 SRM. The stator phases are excited sequentially with a total time of 25 ms required to excite all three phases. Find the angular velocity of the rotor. Express your answer in rad/s and rev/m.

6.3

The following flux equation describes the nonlinear characteristics of a three-phase, 6/4 SRM:

$$\lambda_j(i,\theta) = \lambda_s\left(1 - \exp\left(-i_j f_j(\theta)\right)\right), \quad i_j \geq 0$$

where

$$\lambda_s = \text{saturation flux} = .2\,V - s$$

and

$$f(\theta) = a + b * \cos\left(N_r\theta - (j-1)2\pi/m\right)$$

Here, $j = 1,2,3$ denotes the phase number, and $m = 3$. Also, $a = 0.024$ and $b = 0.019$.

Derive the expression for the phase torque $T_j(i,\theta)$.

7 Power Electronics and Motor Drives

The electric motor drive converts the stiff DC battery voltage to a DC (for DC motor) or an AC (for AC motor) voltage with a root mean square (RMS) value and frequency that can be adjusted according to the control command. The driver input command is translated into a torque command for the motor drive. The torque command, in conjunction with the feedback signals from sensors, sets the operating point parameters for the electric motor and accordingly controls the turn-on and turn-off of the power switches inside the drive system. The motor drive then delivers power at the desired voltage and frequency to the motor, which in turn delivers the desired torque and speed for propulsion.

7.1 ELECTRIC DRIVE COMPONENTS

A motor drive consists of a power electronic converter and the associated controller. The power electronic converter is made of solid-state devices and handles the flow of bulk power from the source to the motor input terminals. The controller processes the command input and the feedback information to generate the switching signals for the power converter semiconductor switches. The coordination of the components of the motor drive with the source and the motor are shown schematically in Figure 7.1.

7.1.1 POWER CONVERTERS

The power converter can be a DC drive supplying a DC motor or an AC drive supplying an AC motor. The converter functions of the two types of drives are shown in Figure 7.2. The power electronic drive requirement for switched reluctance (SR) machines is different from that of DC or AC drives and, hence, will be discussed separately in Chapter 8.

A power converter is made of high-power fast-acting semiconductor devices, such as bipolar junction transistor (BJT), metal oxide semiconductor field effect transistor (MOSFET), insulated gate bipolar transistor (IGBT), silicon-controlled rectifier or thyristor (SCR), gate turn-off SCR (GTO), and MOS-controlled thyristor (MCT). These solid-state devices configured in a certain circuit topology function as an on–off electronic switch to convert the fixed supply voltage into variable voltage and variable frequency supply. All of these devices have a control input gate or base through which the devices are turned on and off according to the command generated by the controller. Tremendous advances in power semiconductor technology over the past two decades enabled the development of compact, efficient, and reliable DC-DC and DC-AC power electronic converter topologies.

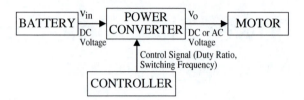

FIGURE 7.1 Block diagram of a motor drive.

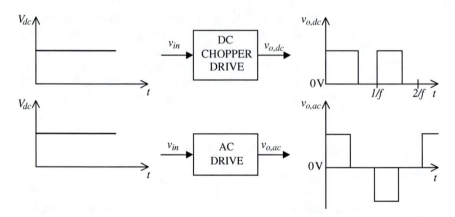

FIGURE 7.2 DC-DC and DC-AC converter functions.

7.1.2 DRIVE CONTROLLER

The drive controller runs the motor drive subsystem of the electric or hybrid electric vehicle. The drive controller is a local controller within the vehicle control system. The vehicle controller sends torque commands to the drive controller, which attempts to deliver the desired request using its internal control algorithm. The drive controller manages and processes the information of the system to control the flow of power into the drivetrain. The function of the drive controller is to accept command and feedback signals, process the information according to the desired criterion, such as efficiency maximimzation, and generate the gate switching signals for the power devices of the converter. Modern controllers are digital instead of analog, which helps minimize drift and error and improves performance through their capability of processing complex algorithms in a short time. The controllers are essentially an embedded system, where microprocessors and digital signal processors are used for signal processing. Interface circuits consisting of A/D and D/A converters are required for communication between the processor and the other components of the system.

7.2 POWER ELECTRONIC SWITCHES

An electronic switch is one that can change an electric circuit configuration by switching states from on to off and vice versa. The electronic switch of Figure 7.3a

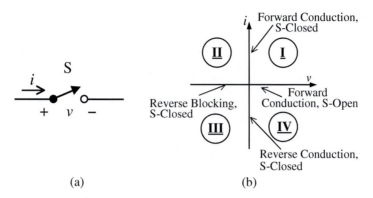

(a)   (b)

FIGURE 7.3 Ideal switch and its conduction characteristics. (a) Switch symbol; (b) switch characteristics in four quadrants.

is an ideal four-quadrant switch, which means that it can handle bidirectional current as well as bidirectional voltage without any power loss. In the ideal switch, there is no voltage drop across the device when it conducts. The ideal switch also turns on and off instantaneously. The quadrants are labeled in Figure 7.3b in the counterclockwise direction of the i-v plane of switch conduction characteristics. Ideally, the operating point will be along the axes, with the ideal switches carrying current with zero voltage drop across it or blocking a voltage with zero current flowing through it. The practical semiconductor switch differs from the ideal switch with respect to conduction voltage drop and finite time required to turn on and turn off. Furthermore, practical devices cannot give four-quadrant capability unless combined with one other device.

Each of the practical devices used in power converters has specific characteristics.[1,2] The particular application and its power requirements determine the type of device to be used in a converter topology. BJTs have higher power ratings and excellent conduction characteristics, but the base drive circuit is complicated, because these are current-driven devices. On the other hand, MOSFETs are voltage-driven devices and, hence, the gate drive circuits are much simpler. The switching frequency of a MOSFET is much higher compared to a BJT, but the maximum available device power ratings would be much smaller for the former. The IGBT is a device, invented in the early 1980s, that combines the positive features of MOSFET and BJT. IGBTs are the devices of choice today, in most cases, due to their availability in high power ratings. Previously, when DC motors were the primary machine choice, high-power converters were typically made of SCRs, which are available in very high power ratings. However, unlike the other devices, SCRs cannot be turned off through a gate signal, and a commutation circuit is required to turn them off. The GTOs are a kind of SCR that can be turned off through a gate signal, although the current required in the gate signal to turn them off is typically four to five times the current required to turn them on. Attempts to combine the gating characteristics of a MOSFET and the conduction characteristics of an SCR resulted in devices called MCTs in the late 1980s and early 1990s. However, failure of the MCTs under certain conditions did not make these devices popular. In addition to the switches mentioned above, there is an additional two-terminal device called a diode, which is universally used in all converters. Diodes are used in conjuction with other controlled devices in the power converter to provide current paths for inductive circuits or for blocking reverse voltages. The important features of the devices discussed above are summarized in Table 7.1. Further details of the operating characteristics of the devices that are significant in EV and HEV applications are discussed in the following.

7.2.1 DIODE

The diode is a two-terminal, uncontrollable switch, that is turned on and off by the circuit. A positive voltage across the anode and the cathode of the diode turns the device on, allowing current conduction up to its rated value. There will be a small forward voltage drop during diode conduction, as shown in Figure 7.4. The diode conducts current in one direction only and blocks voltage in the negative direction, which makes it a Quadrant II switch. The diode can block a reverse voltage to its breakdown level.

7.2.2 POWER TRANSISTORS

The power bipolar junction transistor (BJT) is a three-terminal controlled switch. The circuit symbol and the $i\text{-}v$ characteristics of a BJT are shown in Figure 7.5. When sufficient positive base current i_B flows through the base of a npn BJT, the transistor action allows large positive collector current i_C to flow through the junctions of the device with a small positive collector to emitter voltage v_{CE} (saturation) drop. The amplitude of the base current determines the amplification in the collector current. The power transistors are always operated as a switch, at saturation with high enough base current or at cut-off with zero base current. The power BJT is a controllable switch that can be turned on or off with the help of the base current. The device allows forward current or blocking voltage and, hence, is a Quadrant I switch. Steady state operation in Quadrants II and IV is not possible.

7.2.3 POWER MOSFETs

The power MOSFET is a controlled switch similar to the power BJT. The three terminals of the MOSFET are gate, drain, and source, which are analogous to the base, collector, and emitter, respectively, of the BJT. Similar to the power BJT, the MOSFET is a Quadrant I switch allowing forward flow of current and blocking forward voltage between drain and source. The device works as a controlled switch through control of the voltage between gate and drain. The current-voltage characteristics of a MOSFET are illustrated in Figure 7.6a. Unlike the BJT, the MOSFET is a voltage-driven device with much faster switching capabilities. This allows power MOSFET operation at much higher frequencies compared to the BJT, although the available maximum device voltage and current ratings are much smaller than the BJT.

7.2.4 IGBT

The excellent conduction characteristics of the power BJT were combined with the excellent gate characteristics of the power MOSFET in the insulated gate bipolar

TABLE 7.1
Summary of Power Devices

Name	Symbol	Turn On	Turn Off	Comments
Diode	Cathode ▲ Anode	• Positive anode to cathode voltage	• Reverse anode current • Recovery time before turning off	• Turn off and on depend on circuit conditions • High power capabilities •
SCR	Cathode Gate ▲ Anode	• Small gate pulse (current) • Slow to medium turn-on time (~5 μs)	• Anode current goes below holding • Delay time before forward voltage can be applied (10–200 μs) •	• Very high power • Needs additional circuit to turn off • On voltage ≈2.5 V
GTO	Cathode Gate ▲ Anode	• Small gate pulse (current) • Slow to medium turn-on time (~10 μs)	• Remove charge from gate (medium current) • Medium speed (~0.5 μs)	• High power • Easier to turn off than SCR • On voltage ≈2.5 V
BJT	Collector Base Emitter	• Medium current to base to turn on • Medium speed (0.5 μs)	• Remove current from base • Medium speed (0.2 μs)	• Medium power • Easy to control • Medium drive requirements • On voltage ≈1.5 V
MOSFET	Drain Gate Source	• Voltage to gate (v_{GS}) • Very high speed (0.2 μs)	• Remove voltage from gate • High speed (0.5 μs)	• Low power • Very easy to control • Simple gate drive requirement • High on losses ≈0.1 Ω on resistance
IGBT	Collector Base Emitter	• Voltage to gate (v_{GS}) • High speed (0.4 μs)	• Remove voltage from gate • High speed (~0.7 μs)	• Medium power • Very easy to control • On voltage ≈3.0 V • Combines MOS and BJT technologies

transistor (IGBT). The IGBT has high input impedance like MOSFET and low conduction losses like BJTs. There is no second breakdown problem like BJT in the IGBT. By chip design and structure, the equivalent drain-to-source resistance R_{DS} is controlled to behave like that of a BJT. Conduction characteristics of an IGBT are shown in Figure 7.6b.

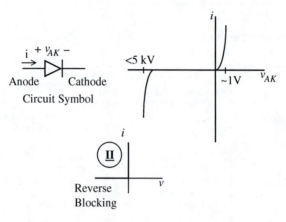

FIGURE 7.4 Diode symbol, characteristics, and operating quadrant.

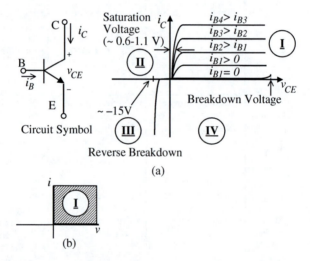

FIGURE 7.5 An npn power BJT: (a) device symbol and *i-v* characteristics; (b) operating quadrant.

7.2.5 BIDIRECTIONAL SWITCH

A positive aspect of an EV or HEV is that the energy during braking of the vehicle need not be wasted in the mechanical brakes through friction but can be recovered to recharge the batteries. Electric motors can operate in the regenerative mode when the vehicle needs to slow down. The direction of power flow is reversed during this mode, with the energy flowing from the wheels to the battery. The regenerative process not only increases the efficiency of the system but also extends the range of the vehicle between recharging of the batteries. Consequently, the power converters need to be bidirectional to recover the energy through regenerative braking. For bidirectional power flow, power devices or switches of the DC-DC or DC-AC

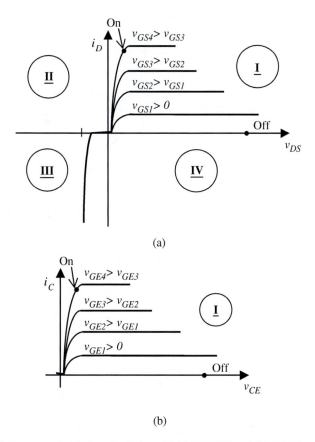

(a)

(b)

FIGURE 7.6 Current-voltage characteristics of (a) MOSFET and (b) IGBT.

converter topologies also need to be bidirectional. Figure 7.7 gives an example of how bidirectional switches can be made by combining a BJT and a diode.

7.3 DC DRIVES

The DC drives for EV and HEV applications are the DC-DC converters, such as DC-choppers, resonant converters, or push–pull converters. A two-quadrant chopper will be analyzed in this chapter as representative of DC drives. The simplicity of the two-quadrant chopper and the torque–speed characteristics of the separately excited DC motor will be utilized to present the interaction of a power electronic motor drive and the vehicle load. Readers interested in the details of the various types of DC drives are referred to the literature.[2–4]

7.3.1 TWO-QUADRANT CHOPPER

The two-quadrant DC chopper allows bidirectional current and power flow with unidirectional voltage supply. The schematic of a two-quadrant chopper is shown

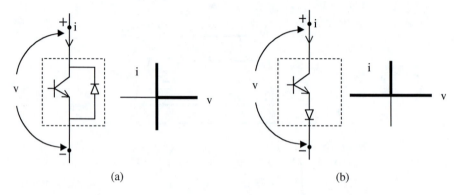

FIGURE 7.7 (a) Bidirectional current switch. (b) Bidirectional voltage switch.

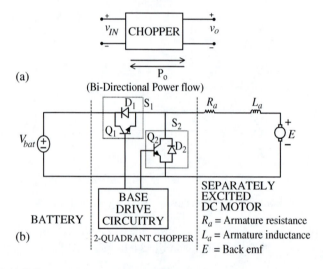

FIGURE 7.8 A DC motor drive. (a) Power flow in drive; (b) drive circuit.

in Figure 7.8. The motor current i_0 is inductive current and, therefore, cannot change instantaneously. The transistor Q_1 and diode D_1 combined make the bidirectional current switch S_1. Similarly, switch S_2 is made of transistor Q_2 and diode D_2. The on and off conditions of the two switches make four switching states (SWS), two of which are allowed and two are restricted, as shown in Table 7.2. In the allowed switching states SWS1 and SWS2, the switches S_1 and S_2 have to withstand positive voltage when they are off and both positive and negative currents when they are on. Therefore, bidirectional current switches have been used.

In Quadrant I operation, turning on Q_1 allows current and power to flow from the battery to the motor. Motor terminal voltage v_0 and current i_0 are greater than or equal to zero. Q_2 is required to remain continuously off in Quadrant I operation and, hence, $i_{b2} = 0$. When Q_1 turns off, D_2 turns on, because i_0 is continuous. Quadrant I operation takes place during the acceleration and constant velocity cruising of a

TABLE 7.2
Switching States of Two-Quadrant Chopper

Switching State	S_1	S_2	Comments
SWS 0	0	0	Not applicable in CCM, because i_0 is inductive
	(OFF)	(OFF)	
SWS 1	0	1	$v_0 = 0$; $v_{S1} = v_{IN}$ (allowed)
		(ON)	$i_{IN} = 0$; $i_{S2} = -i_0$
SWS 2	1	0	$v_0 = v_{IN}$; $v_{S2} = v_{IN}$ (allowed)
			$i_{IN} = i_0$; $i_{S1} = i_0$
SWS 3	1	1	Not allowed, because v_{IN} will get shorted

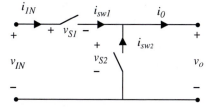 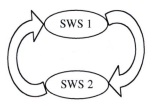

FIGURE 7.9 Quadrant I operation.

vehicle. The chopper operating modes toggle between switching states 1 and 2 in this quadrant, as shown in Figure 7.9.

The transistor Q_1 switches at fixed chopper frequency to maintain the desired current and torque output of the DC motor. The motor current i_0 was shown in Figure 7.10 with exaggerated ripple, where in practice, the ripple magnitude is much smaller compared to the average value of i_0. The filtering requirements set the time period of switching such that a smooth current and torque output is available. The output of the outer-loop vehicle controller desiring a specific torque output of the motor is the duty-ratio d, which sets the on-time of the transistor Q_1. d is a number between 0 and 1, which when multiplied by the time period T, gives the on-time of the transistor. The gate drive signal for Q_1 is a function of d, and consequently, assuming ideal switching conditions, the input voltage to the motor is also dependent on d. The circuit configurations in the two allowed switching states are shown in Figure 7.11 to aid the steady state analysis of the drive system. The steady state analysis is carried out assuming the ideal conditions that there are no switching losses and no delay in the turn-on and turn-off of the devices.

7.3.2 Open-Loop Drive

From a systems perspective, the two-quadrant chopper drives the DC motor that delivers power to the transmission and wheels for vehicle propulsion, as shown in Figure 7.12. Input to the system comes from the driver of the vehicle through the acceleration pedal and the brake pedal (Figure 7.13). Acceleration and constant speed

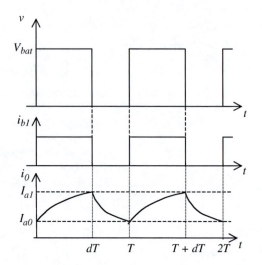

FIGURE 7.10 Output voltage, gate drive for Q_1, and motor current.

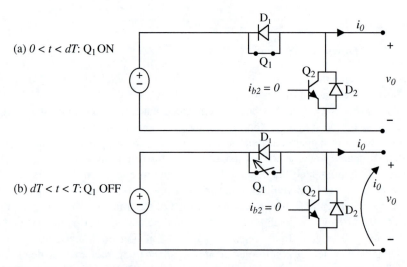

FIGURE 7.11 Circuit condition for (a) switch Q_1 on and (b) for switch Q_2 off.

cruising are controlled by Q_1 in Quadrant I operation, while braking is controlled by Q_2 in Quadrant II operation. In a simplified vehicle control strategy, the slope of the acceleration pedal dictates the desired vehicle motion, and the angle of the pedal is proportionately used to set the duty ratio d_1 for Q_1. Similarly, the slope of the brake pedal expresses the amount of braking desired, and the angle of the brake pedal is proportionately used to set the duty ratio d_2 for Q_2. The two pedals must not be depressed simultaneously.

As mentioned in Chapter 5, one of the advantages of using electric motors for vehicle propulsion is to save energy during vehicle braking through regeneration.

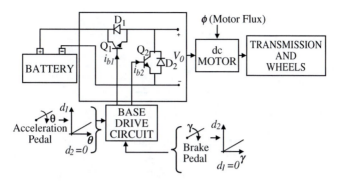

FIGURE 7.12 Open-loop drive for bidirectional power flow.

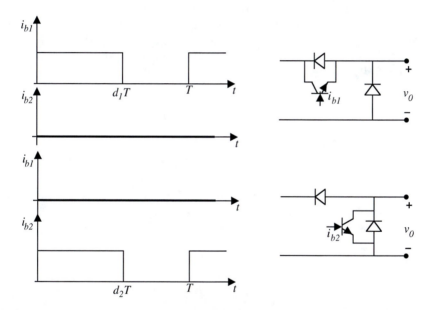

FIGURE 7.13 Base drive signals for the two transistors during acceleration and regeneration.

The energy from the wheels is processed by the power converter and delivered to the battery or other energy storage system during regenerative braking. For the two-quadrant chopper, the amount of regeneration per cycle is a function of the duty ratio, as will be shown later. Therefore, the real duty ratio commands of an EV or HEV will be nonlinearly related to the pedal angle, which is assumed in the simplified analysis to follow. However, the simplistic assumption will give good insight into system control.

7.3.2.1 Steady State Analysis of Quadrant I

During acceleration of the vehicle with the two-quadrant chopper, power and current flow into the motor from the source. The current can be continuous or discontinuous

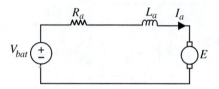

FIGURE 7.14 Equivalent circuit with Q_1 on.

depending on the required torque, although the average value of the current is a nonzero positive value. These two scenarios give the two modes of operation of the two-quadrant chopper, namely, the continuous conduction mode (CCM) and the discontinuous conduction mode (DCM). Let us first analyze the CCM by assuming that the required torque and the chopping frequency to propel the vehicle forward are high enough to maintain a positive current into the motor continuously, as shown in Figure 7.10. The current i_0 is the armature current. Therefore, $i_0 = i_a$. Neglecting ripple in the back-emf E due to ripple in ω, we can assume that E is a constant.

During $0 \leq t \leq dT$, the equivalent circuit of the motor is as shown in Figure 7.14. Applying KVL around the circuit loop,

$$V_{bat} = R_a i_a + L_a \frac{di_a}{dt} + E$$

The initial condition is $i_a(0) = I_{a0} > 0$. This is a first-order linear differential equation, which when solving, one gets

$$i_a(t) = \frac{V_{bat} - E}{R_a}\left(1 - e^{-t/\tau}\right) + I_{a0}e^{-t/\tau}$$

The final condition of SWS2 is

$$i_a(dT) = \frac{V_{bat} - E}{R_a}\left(1 - e^{-dT/\tau}\right) + I_{a0}e^{-dT/\tau} = I_{a1} \tag{7.1}$$

During $dT < t \leq T$, the equivalent circuit of the motor is as shown in Figure 7.15. Let $t' = t - dT$. Applying KVL,

$$0 = R_a I_a + L_a \frac{di_a}{dt} + E$$

Solving the linear differential equation,

$$i_a(t') = -\frac{E}{R_a} + \left(I_{a1} + \frac{E}{R_a}\right)e^{-t'/\tau} = \frac{E}{R_a}\left(-1 + e^{-t'/\tau}\right) + I_{a1}e^{-t'/\tau}$$

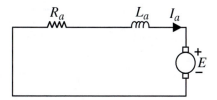

FIGURE 7.15 Equivalent circuit when Q_2 is off.

In steady state, $i_a(t' = t - dT) = I_{a0}$. Therefore,

$$I_{a0} = \frac{E}{R_a}\left(-1 + e^{-T(1-d)/\tau}\right) + I_{a1}e^{-T(1-d)/\tau} \qquad (7.2)$$

Using Equations 7.1 and 7.2 to solve for I_{a0} and I_{a1}, we get

$$I_{a0} = \frac{V_{bat}}{R_a}\left(\frac{e^{dT/\tau} - 1}{e^{T/\tau} - 1}\right) - \frac{E}{R_a}$$

$$I_{a1} = \frac{V_{bat}}{R_a}\left(\frac{1 - e^{-dT/\tau}}{1 - e^{-T/\tau}}\right) - \frac{E}{R_a}$$

The armature current ripple is

$$\Delta i_a = I_{a1} - I_{a0}$$

$$= \frac{V_{bat}}{R_a}\left[\frac{1 + e^{T/\tau} - e^{-dT/\tau} - e^{(1-d)T/\tau}}{e^{T/\tau} - 1}\right] \qquad (7.3)$$

If the armature current i_a has ripple, motor torque T_e will have ripple, because the motor torque is proportional to the armature current ($T_e = K\Phi i_a$). The speed is also proportional to the electromagnetic torque (see Equation 5.4 in Chapter 5). Therefore, for EV and HEV applications, significant ripple in T_e is undesirable, because ripple in torque causes ripple in speed ω, resulting in a bumpy ride. For a smooth ride, ripple in T_e needs to be reduced.

7.3.2.2 Ripple Reduction in i_a

The armature current ripple can be reduced in one of two ways:

1. Adding a series armature resistance.
2. Increasing the chopper switching frequency.

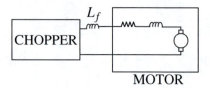

FIGURE 7.16 Series inductance L_f added in series with the chopper.

Adding a series inductance in the armature increases the electric time constant τ. The new time-constant is

$$\tau = \frac{L_f + L_a}{R_a}$$

where L_f is the added series inductance (Figure 7.16). As τ increases, Δi_a decreases for fixed switching period T. The trade-off is the increase in i^2R losses due to L_f, because practical inductance has series resistance. Also, the electrical response time will increase due to the increase in the time-constant.

Increasing the switching frequency of the chopper, i.e., decreasing T, will also reduce the armature current ripple. The upper limit on switching frequency depends on type of switch used. The switching frequency must also be smaller than the controller computational cycle time, which depends on the speed of the processor used and the complexity of the control algorithm. The trade-off of using a higher switching frequency is higher switching losses in the power devices.

7.3.2.3 Acceleration (Continuous Conduction Mode, CCM)

In the acceleration mode, current and power are flowing into the motor from the battery or energy source. The condition for CCM mode is

$$I_{a0} > 0 \Rightarrow I_{a0} = \frac{V_{bat}}{R_a}\left(\frac{e^{d_1 T/\tau} - 1}{e^{T/\tau} - 1}\right) - \frac{E}{R_a} > 0$$

$$\Rightarrow V_{bat}\left(\frac{e^{d_1 T/\tau} - 1}{e^{T/\tau} - 1}\right) > E$$

Note that

$$0 \leq \frac{e^{d_1 T/\tau} - 1}{e^{T/\tau} - 1} \leq 1$$

because $0 \leq d_1 \leq 1$. It follows that

$$V_{bat} > V_{bat} \left(\frac{e^{d_1 T/\tau} - 1}{e^{T/\tau} - 1} \right)$$

Therefore, the condition for CCM is

$$V_{bat} \geq V_{bat} \left(\frac{e^{d_1 T/\tau} - 1}{e^{T/\tau} - 1} \right) > E \qquad (7.4)$$

The electrical time constant of the power converter is much faster than the mechanical time constant of the motor and the vehicle. The analysis of the interaction between the motor torque–speed characteristics and the vehicle force-velocity characteristics is best conducted on an average basis over one time period. The KVL around the motor aramature circuit loop is

$$v_a(t) = R_a i_a(t) + L_a \frac{di_a}{dt} + K\phi\omega(t)$$

Averaging both sides yields

$$\langle v_a \rangle = R_a \langle i_a \rangle + K\phi \langle \omega \rangle \qquad (7.5)$$

The average armature circuit can be represented by the circuit in Figure 7.17. The average armature voltage in the CCM is

$$\langle v_a \rangle = \frac{1}{T} \int_0^T v_a(\tau) d\tau = \frac{V_{bat} d_1 T}{T} = d_1 V_{bat} \qquad (7.6)$$

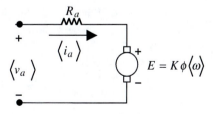

FIGURE 7.17 Average armature circuit.

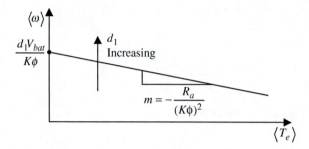

FIGURE 7.18 Torque–speed characteristics of the chopper-fed DC drive.

The average torque equation is

$$T_e(t) = K\phi i_a(t)$$

$$\Rightarrow \langle i_a \rangle = \frac{\langle T_e \rangle}{K\phi}$$

Substituting the average current in Equation 7.6 yields the following:

$$d_1 V_{bat} = R_a \frac{\langle T_e \rangle}{K\phi} + K\phi \langle \omega \rangle$$

$$\Rightarrow \langle \omega \rangle = \frac{d_1 V_{bat}}{K\phi} - \frac{R_a}{(K\phi)^2} \langle T_e \rangle$$

(7.7)

The average torque-speed characteristic of a separately excited DC motor driven by a two-quadrant chopper in the CCM given by Equation 7.6 is shown qualitatively in Figure 7.18. The effect of increasing the duty-ratio d_1 in the acceleration mode is to shift the no-load speed and the rest of the characteristics vertically upwards in the first quadrant.

7.3.2.4 Acceleration (Discontinuous Conduction Mode, DCM)

When the torque required from the motor in the acceleration mode is not high enough, the chopper may enter the DCM, where the armature current becomes discontinuous, as shown in Figure 7.19.

In the discontinuous conduction mode, $I_{a0} \leq 0$ and $V_{bat} > E$ (because power is still flowing from the energy source into the motor). The operation is still in Quadrant I. The condition for DCM of operation is

$$V_{bat} \left(\frac{e^{d_1 T / \tau} - 1}{e^{T / \tau} - 1} \right) \leq E < V_{bat}$$

(7.8)

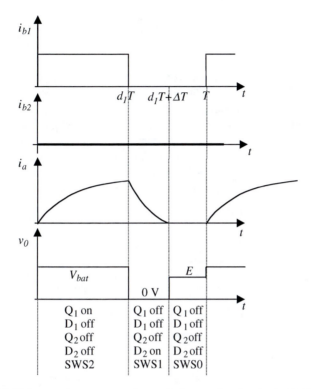

FIGURE 7.19 Voltage and current waveforms during acceleration in the DCM.

The motor armature current cannot go negative without the use of switch Q_2 in the DCM. Q_1 is used to control the flow of power from the source to the motor only. In the interval between $d_1 T + \Delta T$ and T, for i_a to become negative, D_1 must turn on. However, during this interval $v_0 = V_{bat} > E$; hence, i_a cannot become negative. Also, i_a cannot become positive, because that would require $v_0 = V_{bat}$, which would require Q_1 to be on, but $i_{b1} = 0$. Hence, $i_a = 0$ over this interval.

7.3.2.5 Acceleration (Uncontrollable Mode, UNCM)

When a vehicle is rolling down a steep slope, it is possible for the propulsion motor to attain a large value of back-emf. In such a case, if $E > V_{bat}$, current cannot be forced into the motor, and the use of Q_1 becomes meaningless. The supply voltage saturation limit prevents the driver from supplying more power into the motor to move faster than the velocity attained due to gravity. Therefore, the driver cannot control the vehicle using the acceleration pedal; he can only slow the vehicle by using the brake pedal. If the brake pedal is not used in this situation, then the vehicle enters the uncontrollable mode. When $E > V_{bat}$, i_a starts to decrease, and once it reaches zero, diode D_1 becomes forward biased and turns on. i_a continues to increase in the negative direction until it reaches its steady state value of

$$i_a = \frac{-E + V_{bat}}{R_a} \qquad (7.9)$$

The mode of operation in this stage is in Quadrant II. The current and switch conditions are shown in Figure 7.20. Depressing the acceleration pedal to increase d_1 does not control the vehicle in any way, and the vehicle is, in fact, rolling downwards while regenerating into the source in an uncontrolled way. The protection mechanism must kick in at this stage to prevent overcharging of the batteries. Of course, the driver can regain control by switching to the brake pedal from the acceleration pedal and forcing controlled regeneration through the use of Q_2. This will help slow the vehicle on a downhill slope.

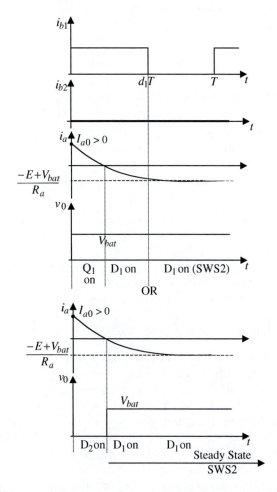

FIGURE 7.20 Voltage and current waveforms during acceleration in the UNCM.

7.3.2.6 Braking Operation (CCM in Steady State)

The most efficient way of recovering energy during vehicle braking is through regeneration in the motor drive system. Let us assume that $E < V_{bat}$, and the brake pedal is depressed. Q_1 is kept off during this period, while braking is controlled through the gate signal i_{b2}. For regeneration, the power flow must be from the motor to the energy source storage, requiring armature current i_a to be negative. Turning Q_2 on helps i_a become negative (from a previously positive value), and an average negative current can be established in a relatively short time for vehicle braking and regeneration. Voltage and current waveforms during braking operation in CCM are shown in Figure 7.21.

An analysis similar to the CCM during acceleration will yield the I_{a1} and I_{a2} values in the steady state CCM during braking as

$$I_{a1} = \frac{1}{R_a} \left\{ V_{bat} \left(\frac{1 - e^{-d_2' T/\tau}}{1 - e^{-T/\tau}} \right) - E \right\}, d_2' = 1 - d_2$$

(7.10)

$$I_{a2} = \frac{1}{R_a} \left\{ V_{bat} \left(\frac{e^{d_2' T/\tau} - 1}{e^{T/\tau} - 1} \right) - E \right\} < 0$$

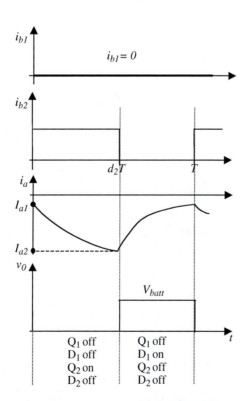

FIGURE 7.21 Voltage and current waveforms during braking operation in CCM.

The current ripple during braking is

$$\Delta i_a = \frac{V_{bat}}{R_a}\left[\frac{-e^{-d_2'T/\tau}+1+e^{T/\tau}-e^{d_2T/\tau}}{e^{T/\tau}-1}\right] \tag{7.11}$$

During braking CCM, $I_{a1} < 0$ and $E < V_{bat}$. Also, note that

$$0 < \left(\frac{1-e^{-d_2'T/\tau}}{1-e^{-T/\tau}}\right) = \frac{e^{T/\tau}-e^{d_2T/\tau}}{e^{T/\tau}-1} < 1$$

Therefore, the condition for continuous conduction during braking is

$$0 < V_{bat}\left(\frac{1-e^{-d_2'T/\tau}}{1-e^{-T/\tau}}\right) < E < V_{bat} \tag{7.12}$$

The average voltage equation in the braking CCM is

$$\langle v_a \rangle + R_a \langle i_a \rangle + 0 = E = K\phi\langle\omega\rangle \tag{7.13}$$

The average motor torque is

$$\langle T_e \rangle = -K\phi\langle i_a \rangle \tag{7.14}$$

The average motor terminal voltage is

$$\langle v_a \rangle = \frac{1}{T}\int_0^T v_a(\tau)d\tau = \frac{1}{T}V_{bat}(T-d_2T) = (1-d_2)V_{bat} \tag{7.15}$$

Substituting in $\langle v_a \rangle$ from Equation 7.15 and $\langle i_a \rangle$ from Equation 7.14 into Equation 7.13,

$$(1-d_2)V_{bat} - \frac{R_a}{K\phi}\langle T_e \rangle = K\phi\langle\omega\rangle$$

The average speed-torque characteristics are as follows (Figure 7.22):

$$\langle\omega\rangle = \frac{(1-d_2)V_{bat}}{K\phi} - \frac{R_a}{(K\phi)^2}\langle T_e \rangle \quad \text{for} \quad \langle T_e \rangle < 0, \langle\omega\rangle > 0 \tag{7.16}$$

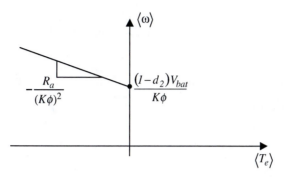

FIGURE 7.22 DC motor drive speed–torque characteristics during braking.

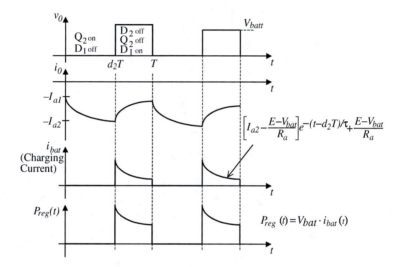

FIGURE 7.23 Voltage, current, and power waveforms during regenerative braking.

7.3.2.7 Regenerative Power

Power is regenerated into the energy source only during part of the cycle, when current flows into the battery from the motor, as shown in Figure 7.23. When transistor Q_2 is on, power is only being dissipated in the switch and contact resistances. When Q_2 is off and diode D_1 is conducting, the current termed $i_{bat}(t)$ is flowing into the battery. Therefore, the instantaneous regenerative power is $P_{reg}(t) = V_{bat} \times i_{bat}(t)$. The average regenerative power is

$$\langle P_{reg} \rangle = \frac{1}{T} \int_{d_2 T}^{T} P_{reg}(\gamma)\, d\gamma \qquad (7.17)$$

Using an analysis similar to that done for the CCM during acceleration and the results of Equation 7.10, the battery current can be derived as

$$
i_{bat}(t) = \begin{cases} \left[I_{a2} - \dfrac{E - V_{bat}}{R_a} \right] e^{-(t - d_2 T)/\tau} + \dfrac{E - V_{bat}}{R_a} & \text{for } d_2 T < t \le T \\ 0 & \text{otherwise} \end{cases} \tag{7.18}
$$

The average regenerative power obtained by inserting Equation 7.18 into Equation 7.17 and integrating is

$$
\langle P_{reg} \rangle = \frac{V_{bat}^2}{R_a} \left[\left(\frac{E}{V_{bat}} - 1 \right)(1 - d_2) + \frac{\tau}{T} \left\{ \frac{e^{(1-d_2)T/\tau} + e^{d_2 T/\tau} - 1}{1 - e^{T/\tau}} \right\} \right] \tag{7.19}
$$

The regenerative energy per cycle is

$$
\int_0^T P_{reg}(\gamma)\,d\gamma = \int_{d_2 T}^T P_{reg}(\gamma)\,d\gamma = \langle P_{reg} \rangle T
$$

7.4 OPERATING POINT ANALYSIS

The following discussion presents the steady state operating point analysis of the vehicle system at the intersection of the motor torque–speed characteristics with the road load characteristics. Four operating points are chosen for analysis in three different quadrants of the motor speed–torque plane representing four chopper operating modes discussed earlier. These modes are acceleration CCM in Quadrant I (Scenario 1), acceleration CCM in Quadrant IV (Scenario 2), acceleration UNCM in Quadrant II (Scenario 3), and braking CCM in Quadrant II (Scenario 4). The four scenarios are shown in Figure 7.24.

Now, recall the tractive force vs. vehicle speed characteristics of Chapter 2, which essentially define the speed–torque characteristics of the EV or HEV load. The tractive force vs. velocity characteristics of Figure 2.11 can be converted into equivalent vehicle load speed–torque characteristics, knowing the transmission gear ratio and the vehicle wheel radius. The steady-state operating point of the vehicle for certain conditions of the motor drive system and road load characteristics can be obtained by overlaying the two speed–torque characteristics on the same plot. The steady-state operating points at the intersection of the motor and load characteristics are shown in Figure 7.25.

7.4.1 SCENARIO 1

In this scenario, the vehicle is moving forward on a level roadway with a constant velocity. The chopper is in the acceleration CCM of operation.

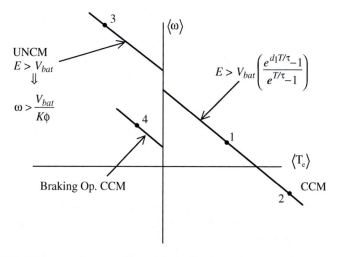

FIGURE 7.24 Motor speed-torque characteristics for four chopper modes.

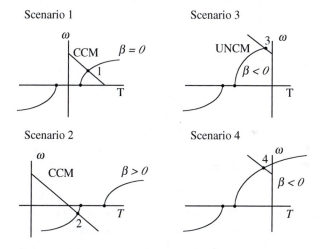

FIGURE 7.25 Operating point analysis for the scenarios.

7.4.2 Scenario 2

The chopper is operating in the acceleration CCM, yet the vehicle is moving backward on a steep uphill road. If the duty ratio has not yet reached 100%, d_1 can be increased all the way up to 1 by depressing the acceleration pedal further and increasing the torque output of the motor. Increasing d_1 will raise the motor speed-torque characteristics vertically upwards, enabling a possible steady state operating point in the first quadrant. If the motor rating has reached its limit, then the torque cannot be increased further to overcome the road load resistance. The slope is too steep for the rating of the motor, and the vehicle rolls backward. The wheel back-drives the motor in this case.

7.4.3 SCENARIO 3

The vehicle is going downhill, with the chopper operating in the acceleration mode. The angle of the acceleration pedal will have no bearing on the steady state operating point. There will be uncontrolled regeneration into the energy source.

7.4.4 SCENARIO 4

The vehicle is going downhill in a controlled fashion using the brake pedal. The speed of the vehicle going down the slope is in the control of the driver.

EXAMPLE 7.1

An EV's drivetrain with a 72 V battery pack is shown in Figure 7.26. The duty ratio for the acceleration operation is d_1, while the duty ratio for braking operation is d_2. The various parameters are given below:

EV parameters:
$m = 1000$ kg, $C_D = 0.2$, $A_F = 2$ m², $C_0 = 0.009$, $C_1 = 0$
$\rho = 1.1614$ kg/m³, and $g = 9.81$ m/s²
r_{wh} = radius of wheel = 11 in = 0.28 m

Motor and controller parameters:
Rated armature voltage $V_{arated} = 72$ V
Rated armature current $I_{arated} = 400$ A
$R_a = 0.5$ Ω, $L_a = 8$ mH, $K\phi = 0.7$ V-s
f_s = chopper switching frequency = 400 Hz

(a) Find the series filter inductance L_f so that worst-case motor armature current ripple is 1% of I_{arated}.
(b) The vehicle road load characteristic on a level road is $T_{TR} = 24.7 + 0.0051\,\omega_{wh}^2$. What is the EV steady state speed on a level road if $d_1 = 0.7$? Assume an overall gear ratio of one for the transmission system. Find the chopper mode of conduction. [**Scenario 1**]

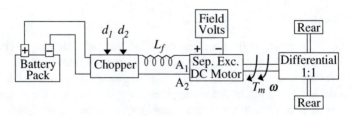

FIGURE 7.26 Two-quadrant chopper-driven DC motor drive for EVs.

(c) What is the percent ripple in the armature current for the operating point in (b)?

(d) The vehicle road load characteristic on a grade of 5.24% ($\beta = 3°$) is T_{TR} = 119.1 + 0.0051 ω_{wh}^2. What is the EV steady state speed for $d_1 = 0.7$? **[Scenario 2]**

(e) What is the EV speed on grade of -5.24% using a brake pedal, with d_2 = 0.5? **[Scenario 4]**

Solution

(a) From Equation 7.3, with $x = e^{T_p/\tau}$, $x > 1$

$$\Delta i_a = I_{a1} - I_{a0}$$

$$= \frac{V_{bat}}{R_a}\left[\frac{1+x-x^{d_1}-x^{(1-d_1)}}{x-1}\right]$$

The worst-case condition is when $d_1 = 0.5$. Therefore,

$$.01*400 = \frac{72}{.5}\left[\frac{1+x-x^{1/2}-x^{1/2}}{x-1}\right] \Rightarrow 1.0278+.972x-2\sqrt{x}=0$$

Solving for x, we get $x = 1$, 1.118. Because $x > 1$, take $x = 1.118$. Therefore,

$$1.118 = e^{T_p/\tau} \Rightarrow \tau = 8.96 \ T_p = 22.4 \text{ ms}$$

Now,

$$\tau = \frac{L_a + L_f}{R_a} = 22.4 \times 10^{-3}$$

$$\Rightarrow L_f = 3.2 \text{ mH}$$

(b) The motor steady state torque–speed characteristics using Equation 7.7 are,

$$<\omega_m> = \frac{.7 \times 72}{.7} - \frac{.5}{.7^2}<T_e> \Rightarrow <\omega_m> = 72-1.02<T_e>$$

The steady state operating point is the point of intersection of the motor torque–speed characteristic and the vehicle road load characteristic T_{TR} = 24.7 + 0.0051 ω_{wh}^2. Solving the two equations, the operating point is $T^* = 32.4$ N, and $\omega^* = 38.9$ rad/s.

Now,

$$V_{bat}\left(\frac{e^{d_1 T_p/\tau} - 1}{e^{T_p/\tau} - 1}\right) = 72 \cdot \frac{e^{.7*2.5/22.4}}{e^{2.5/22.4} - 1} = 49.6$$

And, $E = K\phi <\omega> = 0.7*(38.9) = 27.2$.
Therefore,

$$V_{bat}\left(\frac{e^{d_1 T_p/\tau} - 1}{e^{T/\tau} - 1}\right) > E$$

and, hence, the chopper is operating in CCM. It can also be shown that $I_{a0} > 0$, verifying that the operation is in CCM.

(c)   The ripple in the armature current for the operating point in (b) is

$$\Delta i_a = \frac{72}{.5}\left[\frac{1 + 1.118 - 1.118^7 - 1.118^{(1-.7)}}{1.118 - 1}\right] = 3.37\ A$$

The ripple current magnitude is still less than 1% of rated current.

$$<i_a> = \frac{<T^*>}{K\phi} = \frac{32.4}{.7} = 46.3\ A$$

Therefore, the percentage ripple is

$$\% \text{ripple} = \frac{\Delta i_a}{<i_a>} \times 100\% = \frac{3.37}{46.7} \times 100\% = 7.28\%$$

(d)   The steady state operating point for a grade of –5.24% is at the intersection of the motor torque–speed characteristic $<\omega_m> = 72 - 1.02 <T_e>$ and the vehicle road load characteristic $T_{TR} = 119.1 + 0.0051\ \omega_{wh}^2$. The two solutions are $\omega^* = -40.8, 233$. The correct solution is $\omega^* = -40.8$, because the other speed is too high. The vehicle in this case is actually rolling backward due to insufficient power from the propulsion unit. It can be verified that the maximum gradability of the vehicle is 2.57%.

(e)   The vehicle road load characteristic for a slope of –5.24% is $T_{TR} = -119.1 + 0.0051\ \omega_{wh}^2$. The motor torque–speed characteristic in the braking CCM from Equation 7.14 is

$$\langle\omega_m\rangle = \frac{(1 - d_2)V_{bat}}{K\phi} - \frac{R_a}{(K\phi)}\langle T_e\rangle \Rightarrow \langle\omega_m\rangle = 51.4 - 1.02\langle T_e\rangle$$

Solving for the operating point from the two torque–speed characteristics, $T^* = -57.4, 347$. The negative value is the solution, because the chopper is in braking mode. This gives the steady state speed as $\omega^* = 51.4 -$ $1.02(-57.4) = 109.9$ rad/s. The vehicle speed is $V^* = 0.28 * 109.9 = 30.77$ m/s = 68.8 mi/h.

Now,

$$E = .7 * 109.9 = 76.9 \geq 72 \frac{1 - e^{-.5*.1116}}{1 - e^{-.1116}} = 37$$

which satisfies the condition of CCM for braking given by Equation 7.12.

REFERENCES

1. Baliga, B.J., *Power Semiconductor Devices*, PWS Publishing Company, Boston, MA, 1995.
2. Kassakian, J.G., Schlecht, M.F., and Verghese, G.C., *Principles of Power Electronics*, Addison-Wesley, Reading, MA, 1991.
3. Mohan, N., Undeland, T.M., and Robins, W.P., *Power Electronics: Converters, Applications and Design*, John Wiley & Sons, New York, 1995.
4. Dubey, G.K., *Power Semiconductor Controlled Drives*, Prentice Hall, New York, 1989.

PROBLEMS

Parameters for Problems 7.1 and 7.2 are as follows:

$$V_{bat} = 144 \text{ V}, K\phi = 0.6 \text{ V-s}, L_a = 8E\text{-}4 \text{ H}, R_a = 0.1 \text{ }\Omega, f_s = 500 \text{ Hz}$$

7.1

The time γ (see Figure P7.1) in the discontinuous conduction mode (DCM) during acceleration of a two-quadrant chopper can be derived as

$$\gamma = -\tau \ln\left[\frac{E}{E + V_{bat}\left(e^{d_1 T/\tau} - 1\right)}\right]$$

where

$$\tau = \frac{L_a}{R_a}$$

Derive the $<\omega> - <T>$ characteristics for acceleration operation of the two-quadrant chopper operating in the DCM. (Do not substitute numerical values for

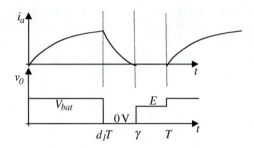

FIGURE P7.1

parameters yet.) Do not try to solve $<\omega>$ in terms of $<T>$. Instead, solve for $<T>$ in terms of $<\omega>$. Plot $<T>$ vs. $<\omega>$ for the given parameters, and $d_1 = 0.9, 0.5,$ and 0.1.

7.2

Calculate the worst-case armature current ripple in CCM for the given parameters. If the worst-case ripple is required to be less than 10 A, what is the value of the filter inductance, or what value should the switching frequency be changed to?

7.3

Find the regions in the T-ω plane for DCM, CCM, and UNCM acceleration operation of a two-quadrant chopper-fed DC motor. That is, find the restrictions on T and ω for each mode. Hint: Start with the condition on E. Solve the inequality for d_1. Then, use the ω-T characteristics to eliminate d_1. Also, remember, $0 \le d_1 \le 1$.

Plot these regions for the given parameters, and also plot the safe operating area given:

$-100 \text{ Nm} \le T \le 100 \text{ Nm}$
$-300 \text{ rad/s} \le \omega \le 300 \text{ rad/s}$
$-30 \text{ hp} \le P \le 30 \text{ hp}$

7.4

Describe the UNCM of braking operation. Draw waveforms of armature current and terminal voltage. Calculate the speed–torque characteristics for this mode. In what quadrant in the ω-T plane is this mode?

7.5

Consider the EV drivetrain driven by a two-quadrant chopper, as shown in Figure 7.26. The duty ratio for the acceleration operation is d_1, while the duty ratio for braking operation is d_2. The various parameters are given below:

EV parameters:

m = 1050 kg, M_B = 150 kg, C_D = 0.25, A_F = 2 m^2, C_0 = 0.01, C_1 = 0
ρ = 1.1614 kg/m^3, and g = 9.81 m/s^2
r_{wh} = radius of wheel = 0.28 m

Motor and controller parameters:

R_a = 0.1 Ω, L_a = 2 mH, $K\phi$ = 0.6 V-s, I_{arated} = 200 A
f_s = chopper switching frequency = 500 Hz
L_f = series filter inductance = 1.6 mH

In each of the following cases, determine whether steady state operation is in CCM, DCM, or UNCM.

(a) d_1 = 0.4, d_2 = 0, V = 25 m/s
(b) d_1 = 0.8, d_2 = 0, V = 45 m/s
(c) d_2 = 0, V = 25 m/s, T = 40 Nm

Note: V is the vehicle steady state velocity, and T is the motor torque. Also, neglect friction and windage loss, and assume zero power loss between the motor shaft and the vehicle wheels.

8 AC and SR Motor Drives

8.1 AC DRIVE

The synchronous speed of AC machines is proportional to the supply frequency ($\omega_s = 120f/P$), which means that speed can be controlled by varying the frequency of the AC input voltage. The power electronic converter supplying the variable voltage, variable frequency output to an AC machine (induction or synchronous) from the available energy source (typically in the form of a fixed DC voltage) is known as the inverter. The inverter consists of six switches, and through appropriate control, it shapes the available DC voltage into balanced three-phase AC voltage of the desired magnitude and frequency. The inverter can be broadly classified as a voltage source inverter or a current source inverter. The voltage source inverter, shown in Figure 8.1, is common for EV and HEV applications, where the source typically delivers a stiff voltage. The six-switch voltage source inverter can operate in the six-step mode or in the pulse width modulation (PWM) mode. The inverter output invariably has a number of harmonic components in addition to the desired fundamental voltage component. PWM is used to minimize the harmonic contents of the output voltage signal. There are several methods of generating PWM signals, such as, sinusoidal PWM, state vector PWM, uniform sampling PWM, selective harmonic elimination, etc. The electronic controller generates the gate switch signals for the inverter power devices using a PWM method and control commands to process the power flow and deliver the output voltage at desired voltage and frequency.

AC Inverter

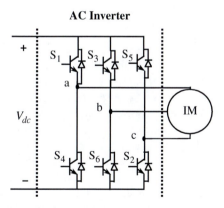

FIGURE 8.1 Six-switch voltage source inverter.

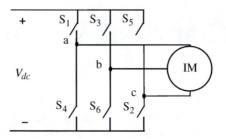

FIGURE 8.2 Ideal six-switch voltage source inverter.

Six-switch inverter topology is used for induction machines as well as for permanent magnet (PM) or any other synchronous machines. This section will discuss inverter operation and control for induction machines, while Section 8.3 will discuss the PM machine drive. The most common method of speed control used in industrial applications is the scalar control method, known as constant volts/hertz control. The method is based on the steady state equivalent circuit of the induction motor, which results in poor performance under dynamic conditions as well as in low-speed and zero-speed operation. The performance requirements of an EV or HEV motor drive are much more rigorous than what can be achieved from a scalar control method; hence, vector control methods based on the reference frame transformation theory are used. Vector control is characterized by involving magnitude and phase control of the applied voltage. Control algorithms are much more complex and computationally intensive for vector control methods. Computational burden is greatly alleviated through the advances of microprocessor and digital signal processor technology. The vector control technique is a well-developed technology with several drive manufacturing companies offering products for various industrial applications.

8.1.1 SIX-STEP OPERATION

The six-step operation is the simplest approach of generating AC voltage using the six-switch inverter. For the sake of analysis, let us replace the transistor and diodes with ideal switches, which results in a simplified equivalent inverter circuit, shown in Figure 8.2. The DC voltage is represented here as V_{dc}. The characteristics of the ideal inverter are that switches can carry current in both directions and the total number of possible switching states is $2^6 = 64$. Some of the switching states are not allowed. For example, S_1 and S_4 cannot be on at the same time. The operation of the inverter can be divided into six intervals between 0 and 2π rad, with each interval being of $\pi/3$ rad duration. In each interval, three of the switches are on, and three are off. This operation is known as the six-step operation. The six interval periods identified with the operating switches during that period are as follows:

$$1 \to 0 < \omega t < \pi/3: \qquad S_1S_5S_6$$
$$2 \to \pi/3 < \omega t < 2\pi/3: \qquad S_1S_2S_6$$
$$3 \to 2\pi/3 < \omega t < \pi: \qquad S_1S_2S_3$$
$$4 \to \pi < \omega t < 4\pi/3: \qquad S_4S_2S_3$$
$$5 \to 4\pi/3 < \omega t < 5\pi/3: \qquad S_4S_5S_3$$
$$6 \to 5\pi/3 < \omega t < 2\pi: \qquad S_4S_5S_6$$

The gating signals for the electronic switches and the resulting output AC voltage for six-step operation are shown in Figure 8.3. The line-to-line voltage and the line-to-neutral voltage, i.e., voltage of one phase, are shown in the figure.

In three-phase machines, three-wire systems are used, where the line terminals a, b, and c are connected to the motor with neutral terminal n kept hidden. The relationship between line-to-line voltages (i.e., line voltages) and line-to-neutral voltages (i.e., phase voltages) are as follows:

$$v_{ab} = v_{an} - v_{bn}$$
$$v_{bc} = v_{bn} - v_{cn} \qquad (8.1)$$
$$v_{ca} = v_{cn} - v_{an}$$

There are three unknowns and two linearly independent equations (v_{an}, v_{bn}, v_{cn}). Therefore, we need another equation to find the inverse solution. For a balanced three-phase electrical system, we know that $v_{an} + v_{bn} + v_{cn} = 0$. Therefore, the line and phase voltage relationships can be written in matrix format:

$$\begin{bmatrix} 1 & -1 & 0 \\ 0 & 1 & -1 \\ -1 & 0 & 1 \\ 1 & 1 & 1 \end{bmatrix} \begin{bmatrix} v_{an} \\ v_{bn} \\ v_{cn} \end{bmatrix} = \begin{bmatrix} v_{ab} \\ v_{bc} \\ v_{ca} \\ 0 \end{bmatrix}$$

The solutions for the phase voltages are as follows:

$$v_{an} = -\frac{1}{3}\left[v_{bc} + 2v_{ca}\right]$$
$$v_{bn} = \frac{1}{3}\left[2v_{bc} - v_{ac}\right] \qquad (8.2)$$
$$v_{cn} = -\frac{1}{3}\left[v_{bc} + v_{ac}\right]$$

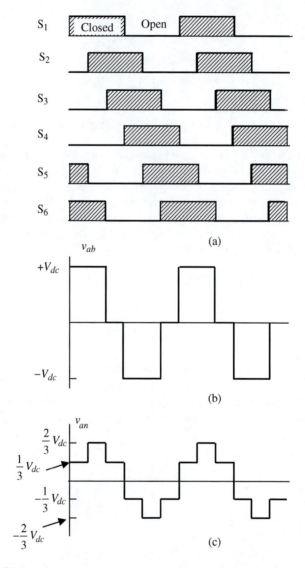

FIGURE 8.3 Six-step inverter gate signals and output voltages. (a) Gate signals for six switches; (b) output line-to-line voltage; (c) output phase voltage.

Phase voltages are useful for per phase analysis of three-phase systems. The phase voltage v_{an} of the six-step inverter output can be derived using Equation 8.2, as shown in Figure 8.3. Now, the question is what type of switches are required for the inverter. Let us first consider switch S_1. When S_1 is off, $v_{S1} = V_{dc}$. When S_1 is on, S_4 is off. The phase voltage of the six-step inverter is shown in Figure 8.4. Output voltage can be filtered to make the supply more sinusoidal. The supply is naturally filtered when supplying an inductive load, such as an electric motor. It is only the fundamental component of the supply that contributes to electromagnetic torque

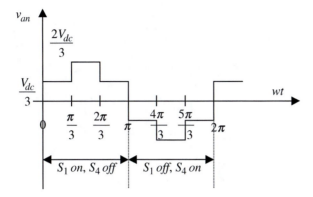

FIGURE 8.4 Phase voltage during switch S_1 operation.

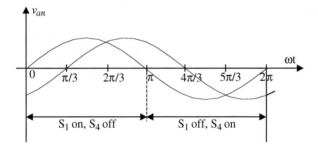

FIGURE 8.5 Phase voltage and current during S_1 operation.

production, while the higher harmonic components are responsible for part of the losses in the system. The interaction of the induction motor with the higher harmonic components is analyzed in the next section. The fundamental component of the phase voltage and the current through the switch into Phase-a winding is shown in Figure 8.5. Because the motor is inductive, line current lags behind phase voltage. Therefore, we must use bidirectional current switches, because i_a is positive and negative when voltage v_{an} is positive.

8.1.1.1 Harmonic Analysis

Let us consider the switching interval number 3, where $2\pi/3 < \omega t < \pi$. The switches that are on during this period are S_1, S_2, and S_3. The inverter configuration in this interval is shown in Figure 8.6. The line-to-line square-wave voltage can be written in terms of the fundamental and harmonic components using Fourier series analysis as follows:

$$v_{ab} = \frac{2\sqrt{3}}{\pi} V_{dc} \left\{ \sin\left(\omega t + \frac{\pi}{6}\right) + \frac{1}{5}\sin\left(5\omega t - \frac{\pi}{6}\right) + \frac{1}{7}\sin\left(7\omega t + \frac{\pi}{6}\right) + \ldots \right\} \quad (8.3)$$

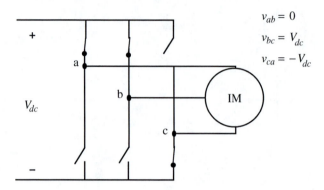

FIGURE 8.6 Inverter switch connection in state number 3.

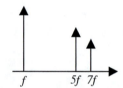

FIGURE 8.7 Dominant harmonic components in inverter output voltages.

Note that the harmonics present are the $6n \pm 1$ (n is an integer) components, and that the triple harmonics are absent. The harmonic phase voltages that are 30° phase shifted from the line voltages are

$$v_{an} = \frac{2}{\pi} V_{dc} \left\{ \sin(\omega t) + \frac{1}{5}\sin(5\omega t) + \frac{1}{7}\sin(7\omega t) + \ldots \right\} \qquad (8.4)$$

The dominant harmonic components are shown in Figure 8.7.

Harmonics do not contribute to the output power, but they increase power losses, which degrades the efficiency and increases the thermal loading of the machine. Harmonic losses do not vary significantly with the load. The interactions of fundamental air gap mmfs with harmonic air gap mmfs produce torque pulsations, which may be significant at low speeds.

8.1.2 PULSE WIDTH MODULATION

Pulse width modulation techniques are used to mitigate the adverse effects of harmonics in the inverter.[1,2] Harmonics in the output PWM voltage are not eliminated but are shifted to a much higher frequency, making filtering a lot easier. Controllability of the amplitude of the fundamental output voltage is another advantage of PWM. Several PWM techniques have been developed over the years, such as sinusoidal PWM, uniform sampling PWM, selective harmonic elimination PWM, space vector PWM, random PWM, etc. The two commonly used PWM techniques, the sinusoidal PWM and the space vector PWM, are discussed in the following.

8.1.2.1 Sinusoidal PWM

In the sinusoidal PWM method, a sinusoidal control signal v_A is compared with a high-frequency triangular waveform v_T to generate the inverter switch gating signals. The frequency of v_T establishes the inverter switching frequency f_c. The magnitude and frequency of the sinusoidal signal are controllable, but the triangular signal magnitude and frequency are kept constant. The sinusoidal control signal v_A modulates the switch duty ratio, and its frequency f is the desired fundamental frequency of the inverter. For the three-phase generation, the same v_T is compared with three sinusoidal control voltages v_A, v_B, and v_C, which are 120° out of phase, with respect to each other, to produce a balanced output. Switches are controlled in pairs (S_1,S_4), (S_2,S_5), and (S_3,S_6). When one switch in a pair is closed, the other switch is open. In practice, there has to be a blanking pulse between the change of control signals for the switches in a pair to ensure that there is no short circuit in the inverter. This is necessary, because practical switches take finite time to turn on and turn off. The three-phase sinusoidal pulse width modulation signals are shown in Figure 8.8. Switch control signals follow the logic given below:

S_1 is on when $v_A > v_T$
S_2 is on when $v_C > v_T$
S_3 is on when $v_B > v_T$
S_4 is on when $v_A < v_T$
S_5 is on when $v_C < v_T$
S_6 is on when $v_B < v_T$

The amplitude modulation index is

$$m = \frac{v_{A,peak}}{v_{T,peak}} = \frac{A}{A_m} \tag{8.5}$$

where A is the amplitude of the reference sine wave, and A_m is the amplitude of the triangular carrier wave. Let us define the ratio of carrier frequency and fundamental frequency as follows:

$$p = \frac{f_C}{f}$$

The rms value of the fundamental inverter output phase voltage v_{ph} is

$$v_{ph,1} = m\frac{V_{dc}}{2\sqrt{2}}, \quad m \le 1 \tag{8.6}$$

where v_{dc} is the DC input voltage to the inverter. Note that $v_{ph,1}$ increases with m until $m = 1$. For $m > 1$, modulation ceases to be sinusoidal PWM wherein it is

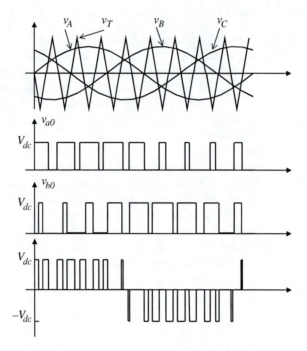

FIGURE 8.8 Three-phase sinusoidal pulse width modulation.

referred to as overmodulation. The rms value of the fundamental in the line-to-line voltage is

$$V_{LL,1} = m\frac{\sqrt{3}V_{dc}}{2\sqrt{2}} \ , \ m \le 1 \tag{8.7}$$

There are two types of modulation depending on the value of p, called synchronous modulation and asynchronous modulation. In synchronous modulation, $p = 3n$, $n = 1,2,...$ making the carrier wave symmetrical with respect to the three-phase reference voltages v_A, v_B, and v_C. Otherwise, the modulation is called asynchronous modulation. The characteristics of asynchronous modulation are as follows:

1. Pulse pattern does not repeat itself identically from cycle to cycle.
2. Subharmonics of f and a DC component are introduced.
3. Subharmonics cause low-frequency torque and speed pulsations known as frequency beats.
4. For large p, frequency beats are negligible. For small p, frequency beats may be significant.
5. For small p, synchronous modulation is used. Preferably, p is an odd multiple of three.

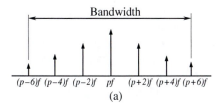
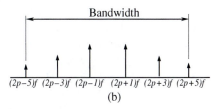

FIGURE 8.9 Harmonic frequency components for (a) $k_1 = 1$, and (b) $k_1 = 2$.

The boundary of sinusoidal modulation is reached when $m = 1$. The relationship between the fundamental component of phase voltage and m ceases to be linear for values greater than one. For sufficiently large values of m, the output phase voltage becomes a square wave with maximum amplitude of the fundamental equal to $2\,V_{dc}/\pi$ (relate to Equation 8.4). This is essentially the six-step operation of the inverter. Note that the amplitude of the fundamental on the boundary of linear sinusoidal PWM is only 78.5% of the maximum value. The modulation with $m > 1$ is known as overmodulation; lower-order harmonics are introduced in this range of modulation.

8.1.2.2 Harmonics in Sinusoidal PWM

The inverter output phase voltage contains harmonics that are of odd multiples of the carrier frequency f_C (i.e., f_C, $3f_C$, $5f_C$, ...). The waveform also contains sidebands centered around multiples of f_C and given by

$$f_h = k_1 f_C \pm k_2 f = \left(k_1 p \pm k_2\right) f$$

where $k_1 + k_2$ is an odd integer. The harmonic frequency centers are at $k_1 f$ for odd integers of k_1, while the sidebands are symmetrically located without a center for even integers of k_1. Note that the magnitudes of band frequency harmonics decrease rapidly with increasing distance from band center. The harmonic frequency components for $k_1 = 1$ and $k_1 = 2$, along with the bandwidth, are shown in Figure 8.9. The bandwidth of harmonics is the range of frequency over which the harmonics are considered dominant. The frequency bandwidth increases with m. The dominant harmonics among the sidebands are at frequencies $(p \pm 2)f$ and $(2p \pm 1)f$.

8.1.2.3 Space Vector PWM

The voltage space vectors embedded in the dq models and vector controllers of AC machines present a highly compatible method for the control of three-phase PWM inverters. The dq voltage commands generated by the controller to drive a motor are converted into an equivalent PWM signal to be applied to the gates of the six inverter switches. For variable frequency drive applications, space vector PWM is popular because of its superior performance compared to other voltage PWM techniques.

The three-phase voltage source inverter can assume one of eight ($2^3 = 8$) states only, because each switch in a phase leg can assume an on or an off position. Each of these eight states is associated with a specific set of values of the three-phase voltages. Based on the state, the machine phase winding terminals are connected to the upper or lower side of the DC link. For terminal voltages, eight voltage vectors are defined using three-bit digital states as 000, 100, 110, 010, 011, 001, 101, and 111. For example, the state 011 stands for phase a connected to the lower side of the DC bus, while b and c phases connected to the upper side. Alternatively, a 0 indicates that the lower switch is turned on, and a 1 indicates that the upper switch is on. This means that the inverter can generate eight stationary voltage vectors V_0 through V_7, with the subscript indicating the corresponding state of the inverter. These voltage vectors are

$$V_0 = \begin{bmatrix} 0 \\ 0 \\ 0 \end{bmatrix} \quad V_1 = \begin{bmatrix} 1 \\ 0 \\ 0 \end{bmatrix} \quad V_2 = \begin{bmatrix} 1 \\ 1 \\ 0 \end{bmatrix} \quad V_3 = \begin{bmatrix} 0 \\ 1 \\ 0 \end{bmatrix} \quad V_4 = \begin{bmatrix} 0 \\ 1 \\ 1 \end{bmatrix} \quad V_5 = \begin{bmatrix} 0 \\ 0 \\ 1 \end{bmatrix}$$

$$V_6 = \begin{bmatrix} 1 \\ 0 \\ 1 \end{bmatrix} \quad V_7 = \begin{bmatrix} 1 \\ 1 \\ 1 \end{bmatrix}$$

The switching states, voltage vectors, and associated switch conditions are given in Table 8.1.

The eight voltage vectors represent two null vectors (V_0 and V_7) and six active state vectors forming a hexagon (Figure 8.10). A simple circuit analysis will reveal that the phase voltage magnitudes vary between $\pm 2/3 V_{dc}$ in steps of $1/3 V_{dc}$ in the six active states. This is analogous to the six-step operation, where voltage magnitude variations are similar for the six steps. For example, in vector state V_1, switches S_1, S_2, and S_6 are on with Phase-a connected to the upper side of the DC bus and Phases b and c connected to the lower side. In this state, the inverter phase voltages are

TABLE 8.1
Space Vector Switching States

Switching State	On Devices	Voltage Vector
0	$S_4 S_6 S_2$	V_0
1	$S_1 S_6 S_2$	V_1
2	$S_1 S_3 S_2$	V_2
3	$S_4 S_3 S_2$	V_3
4	$S_4 S_3 S_5$	V_4
5	$S_4 S_6 S_5$	V_5
6	$S_1 S_6 S_5$	V_6
7	$S_1 S_3 S_5$	V_7

$v_{an} = 2/3V_{dc}$, $v_{bn} = -1/3V_{dc}$, and $v_{cn} = -1/3V_{dc}$. This is the step in the interval $\pi/3 < \omega t < 2\pi/3$ in Figures 8.3 and 8.4 for six-step operation. Transforming the three-phase abc variables into dq variables, the d and q voltages are $v_d = 2/3V_{dc}$ and $v_q = 0$. Therefore, voltage vector $\mathbf{V_1}$ has a magnitude of $2/3V_{dc}$ and is along the d-axis in the dq-plane. It can be shown similarly that each of the six active state vectors has a magnitude of $2/3V_{dc}$ and is displaced by 60° with respect to each other in the dq-plane. The six active state vectors form a hexagon, as shown in Figure 8.10, with the null vectors remaining in the origin. The six active state vectors can be represented in space vector form as

$$\vec{V}_k = \frac{2}{3}V_{dc}e^{j(k-1)\pi/3} \tag{8.8}$$

where $k = 1,\ldots,6$. Note the difference in notation between V_k, which is used to denote the voltage vector in the abc frame strictly in relation to the switch location, and the space vector $\vec{V}_k$, used to define the same vector in mathematical terms.

The space vector (SV) PWM operates in modulated fashion, including the null vectors within a fixed time period, in contrast to the six-step operation of the inverter. The six-step operation uses only the six active states moving in the sequence

$$\mathbf{V_1} \rightarrow \mathbf{V_2} \rightarrow \mathbf{V_3} \rightarrow \mathbf{V_4} \rightarrow \mathbf{V_5} \rightarrow \mathbf{V_6} \rightarrow \mathbf{V_1}$$

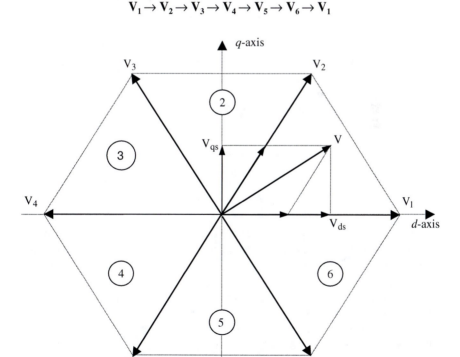

FIGURE 8.10 Inverter switching states and space vectors.

with a duration of 60° in each one of the states. The six-step operation entails increased harmonic distortion. The objective in SV PWM is to generate the gating signals such that harmonically optimized PWM voltage is obtained at the output of the inverter.

8.1.2.4 Generation of SV PWM Switching Signals

The continuous space vector modulation technique is based on the fact that every reference voltage vector $\vec{V}^*$ inside the hexagon can be expressed as a combination of the two adjacent active state vectors and the null state vectors. Therefore, the desired reference vector imposed in each cycle is achieved by switching among four inverter states. The sector where the space vector $\vec{V}^*$ lies determines the two active state vectors that will be used in generating the gate switching signals of a PWM period. The phase angle is evaluated from $\theta = \arctan(V_{qs}/V_{ds})$ and $\theta \in [0, 2\pi]$. The phase angle is related to the relative magnitudes of V_{qs} and V_{ds}. For example, in Sector 1, $0 \leq \arctan(V_{qs}/V_{ds}) < \pi/3$; hence, $0 < V_{qs} < \sqrt{3}\, V_{ds}$. The following conditions are used to determine the sector where the space vector is located:

Sector 1: $0 < V_{qs} < \sqrt{3}\, V_{ds}$

Sector 2: $V_{qs} > \left| \sqrt{3}\, V_{ds} \right|$

Sector 3: $0 < V_{qs} < -\sqrt{3}\, V_{ds}$

Sector 4: $0 > V_{qs} > \sqrt{3}\, V_{ds}$

Sector 5: $V_{qs} < -\left| \sqrt{3}\, V_{ds} \right|$

Sector 6: $0 > V_{qs} > -\sqrt{3}\, V_{ds}$

Let us assume that $\vec{V}_k$ and $\vec{V}_{k+1}$ are the components of the reference vector $\vec{V}^*$ in sector k and in the adjacent active sector $(k + 1)$, respectively. In order to obtain optimum harmonic performance and minimum switching frequency for each of the power devices, the state sequence is arranged such that the transition from one state to the next is performed by switching only one inverter leg. This condition is met if the sequence begins with one zero-state and the inverter poles are toggled until the other null state is reached. To complete the cycle, the sequence is reversed, ending with the first zero-state. If, for instance, the reference vector sits in Sector 1, the state sequence has to be ...0127210..., whereas in Sector 4, it is ...0547450... The central part of the space vector modulation strategy is the computation of the active and zero-state times for each modulation cycle. These are calculated by

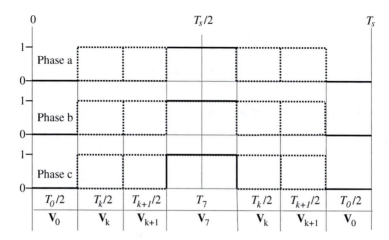

FIGURE 8.11 Diagram of on-times of vectors.

equating the applied average voltage to the desired value. Figure 8.11 demonstrates the on-times of vectors.

In the following, T_k denotes the required on-time of active state vector $\vec{V}_k$, T_{k+1} denotes the required on-time of active state vector $\vec{V}_{k+1}$, and $(T_0 + T_7)$ is the time for the null state vector to cover the complete time period T_s.[3] The on-time is evaluated by the following equation:

$$\int_0^{T_s/2} \vec{V}\,dt = \int_0^{T_0/2} \vec{V}_0\,dt + \int_{T_0/2}^{T_0/2+T_k/2} \vec{V}_k\,dt + \int_{T_0/2+T_k/2}^{T_0/2+T_k/2+T_{k+1}/2} \vec{V}_{k+1}\,dt + \int_{T_0/2+T_k/2+T_{k+1}}^{T_s/2} \vec{V}_7\,dt \quad (8.9)$$

where $T_0 + T_k + T_{k+1} + T_7 = T_s$. By taking $\vec{V}_0$ and $\vec{V}_7$ to be zero vectors, and $\vec{V}^*$, $\vec{V}_k$, and $\vec{V}_{k+1}$ to be constant vectors in a PWM period, Equation 8.9 can be simplified as

$$\frac{T_s}{2}\vec{V} = \frac{T_0}{2}\vec{V}_0 + \frac{T_k}{2}\vec{V}_k + \frac{T_{k+1}}{2}\vec{V}_{k+1} + \frac{T_7}{2}\vec{V}_7$$

$$\Rightarrow T_s\vec{V} = T_k\vec{V}_k + T_{k+1}\vec{V}_{k+1} \quad (8.10)$$

Using the space vector relation for $\vec{V}$ and Equations 8.8 and 8.10, it can be shown that

$$\begin{bmatrix} T_k \\ T_{k+1} \end{bmatrix} = \frac{\sqrt{3}\,T_s}{V_{dc}} \begin{bmatrix} \sin\dfrac{k\pi}{3} & -\cos\dfrac{k\pi}{3} \\ -\sin\dfrac{(k-1)\pi}{3} & \cos\dfrac{(k-1)\pi}{3} \end{bmatrix} \begin{bmatrix} V_{ds} \\ V_{qs} \end{bmatrix} \quad (8.11)$$

The duty ratios are related to the on-time as

$$\begin{bmatrix} T_k \\ T_{k+1} \end{bmatrix} = T_s \begin{bmatrix} D_k \\ D_{k+1} \end{bmatrix} \tag{8.12}$$

Comparing Equations 8.11 and 8.12, one can obtain the duty ratios as

$$\begin{bmatrix} D_k \\ D_{k+1} \end{bmatrix} = \sqrt{3} \begin{bmatrix} \sin\dfrac{k\pi}{3} & -\cos\dfrac{k\pi}{3} \\ -\sin\dfrac{(k-1)\pi}{3} & \cos\dfrac{(k-1)\pi}{3} \end{bmatrix} \begin{bmatrix} \dfrac{V_{ds}}{V_{dc}} \\ \dfrac{V_{qs}}{V_{dc}} \end{bmatrix} \tag{8.13}$$

The on-times for the null vectors are obtained as follows:

$$\begin{bmatrix} T_0 \\ T_7 \end{bmatrix} = \frac{1}{2} \begin{bmatrix} T_s - T_k - T_{k+1} \\ T_s - T_k - T_{k+1} \end{bmatrix} \tag{8.14}$$

Therefore, Equations 8.12 and 8.14 give the on-times T_k, T_{k+1}, T_0, and T_7 with the duty-ratio given by Equation 8.13 within a PWM period. The on-times of the three-phase PWM generator for the controller are determined as follows:

$$\begin{bmatrix} T_A \\ T_B \\ T_C \end{bmatrix} = \begin{bmatrix} \mathbf{V}_k & \mathbf{V}_{k+1} & \mathbf{V}_7 \end{bmatrix} \begin{bmatrix} T_k \\ T_{k+1} \\ T_7 \end{bmatrix} \qquad (k = 1,\dots 6) \tag{8.15}$$

where T_A, T_B, and T_C are the on-times of each phase, and $\mathbf{V}_k$ is one of the six active operating states defined previously. For the instance shown in Figure 8.10, $\vec{V}^*$ is in Sector 1. The on-times are obtained as

$$\begin{bmatrix} T_A \\ T_B \\ T_C \end{bmatrix} = \begin{bmatrix} \mathbf{V}_1 & \mathbf{V}_2 & \mathbf{V}_7 \end{bmatrix} \begin{bmatrix} T_k \\ T_{k+1} \\ T_7 \end{bmatrix} = \begin{bmatrix} 1 & 1 & 1 \\ 0 & 1 & 1 \\ 0 & 0 & 1 \end{bmatrix} \begin{bmatrix} T_k \\ T_{k+1} \\ T_7 \end{bmatrix}$$

The on-times of the switching gates are set up by the SV PWM controller by decomposing the rotating reference vector $\vec{V}^*$ into its constituent dq phase vectors, V_{ds} and V_{qs}. The balanced three-phase voltages are generated in the terminals of the gates.

It was shown that only 78.5% of the inverter's capacity is used with sinusoidal PWM method. In SV PWM method, the inverter's capability is improved by using a separate modulator for each of the inverter legs, generating three reference signals forming a balanced three-phase system. In this way, the maximum obtainable output voltage is increased with ordinary SV PWM up to 90.6% of the inverter capability.[4] The SV PWM algorithm is fairly complex and computation intensive, which limits the PWM switching frequency.

8.1.3 CURRENT CONTROL METHODS

In systems where the output current depends not only on the input voltage but also on the load, a closed-loop PWM method is necessary. The current PWM methods use current feedback information from sensors and generate PWM signals for the inverter gates in a closed-loop fashion. The measured three-phase currents are compared with the three reference current commands generated by the outer-loop controller. The errors between the measurements and the reference signals are utilized in a PWM scheme to generate the gate-switching signals. A number of current controlled PWM methods have been developed, ranging from fairly simple to rather complex. Some of these methods are hysteresis current controller, ramp-comparison controller, predictive current controller, etc. The predictive current controller is one of the more complex, where parameters of the load are used to predict the reference currents. Two of the simpler methods, the hysteresis current controller and the ramp-comparison controller, are described briefly in the following.

8.1.3.1 Hysteresis Current Controller

In the hysteresis current controller, the error between the measured current and the reference current is compared with a hysteresis band, as shown in Figure 8.12a. If the current error is within the band, then PWM output remains unchanged. If the current error exceeds the band, then the PWM output is reversed, forcing a sign change of the slope of di/dt. Mathematically stated, the PWM output is obtained from

$$PWM = \begin{cases} 0 & \text{if} \quad \Delta i < -h/2 \\ 1 & \text{if} \quad \Delta i > h/2 \end{cases}$$

A "0" PWM output signal cuts off the supply voltage in the controlled phase, forcing current to decay, while a "1" PWM output applies voltage to the phase, resulting in an increase of current. The voltage will then force the current to vary in such a way that it remains within the hysteresis band.

The advantage of such a controller is that the error remains within a certain band, which is known by the user. In the hysteresis controller, the switching frequency is unknown, which makes it difficult to design the filters. The switching frequency should be carefully monitored so that the inverter limits are not exceeded. In practical implementations, a frequency limit is used so as not to exceed the inverter maximum switching frequency. The hysteresis band is designed according to the limitations of the device-switching frequency. If the hysteresis band h is chosen to be small, then the switching frequency will be high and not be compatible with the maximum switching frequency of the power devices. On the other hand, if the band is too big, then the current error will be too large.

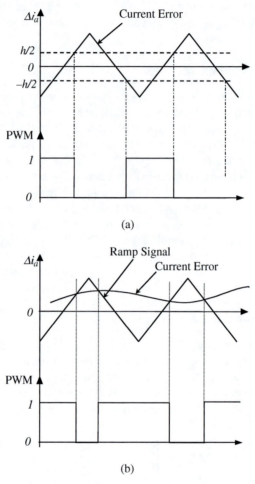

FIGURE 8.12 (a) Hysteresis current control and PWM output. (b) Ramp comparison control and PWM output.

The hysteresis current controller can be used in a three-phase PWM inverter, with each phase having its own PWM controller. If the actual current is higher than the reference current by the amount of half of the hysteresis band, the lower leg switch of the bridge inverter is turned on to reduce the phase current. The difficulty in three-phase hysteresis control is that there may be conflicting requirements of switch conditions for the phases based on the output of the hysteresis controller. The difficulty arises from the interaction of the phases of the three-phase system and the independent hysteresis controllers for each phase. The consequence of this difficulty is that the current does not remain within the hysteresis band. For example, a current-increase command in Phase a needs a return path through Phases b or c lower legs. If Phases b and c happen to have upper leg switches on during this instant, the current in Phase a will not increase to follow the command but will freewheel. In this case, it is possible that the current error of Phase a exceeds the

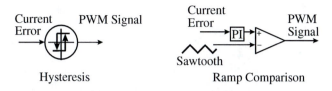

FIGURE 8.13 Hysteresis and ramp comparison techniques.

hysteresis current band. Using the *dq* transformation theory, it is possible to first transform the three-phase currents into two-phase *dq* currents and then impose the hysteresis control in the *dq* reference frame.[1]

8.1.3.2 Ramp Comparison Controller

Another way of controlling the required stator current is to use a controller-based fixed-frequency ramp signal that stabilizes the switching frequency. The current error is first fed into a linear controller, which is typically of the proportional integral (PI) type. The output of the linear controller is compared with a high-frequency sawtooth-shaped triangular signal to generate the PWM switch signals. If the error signal is higher than the triangular signal, the PWM output signal will be "1," and in the other case, the output will be zero. The control actions are shown graphically in Figure 8.12b. Stator voltages will vary in order to minimize the current error signal. Three identical controllers are used in three-phase systems.

The ramp-comparison method has a fixed switching frequency set by the sawtooth wave frequency that makes it easier to ensure that the inverter switching frequency is not exceeded. There are more parameters to adjust in the ramp comparison controller, allowing greater flexibility of control compared to the hysteresis controller. The control parameters include gains of the linear controller, and magnitude and frequency of the sawtooth wave in the case of ramp comparison controller, whereas the only control parameter in a hysteresis controller is the width of the hysteresis band. The functional differences in the two methods are depicted in Figure 8.13.

The primary disadvantage of the ramp comparison controller is increase in response time due to transport delay. The situation can be improved by using a high gain proportional controller instead of a PI controller and by increasing the switching frequency of the sawtooth signal.

8.2 VECTOR CONTROL OF AC MOTORS

EV and HEV propulsion drives require accurate speed control with fast response characteristics. Scalar control methods for induction motors that are easier to implement are inadequate for such applications demanding high performance. The induction motor drive is capable of delivering high performance similar to that of DC motors and PM brushless DC motors using the vector control approach. Although vector control complicates the controller implementation, the lower cost and rugged construction of the induction machines are an advantage over the DC and PM machines.

The key variable for control in speed- and position-controlled applications is the torque. Although torque is never measured directly, torque estimators from machine models are frequently used to generate the current commands. A current controller in the innermost control loop regulates the motor current by comparing the command currents with feedback current measurements coming from the sensors. Speed control, if necessary, is achieved in the outer loop by comparing command speed signal with the feedback speed signal. With the two loops arranged in a cascade, the speed controller output of the outer loop is the current command for the inner loop. In certain high-performance position-controlled applications, such as in the actuator drives for accessories in EVs and conventional vehicles, the position is controlled in the outermost loop, putting the speed controller in an intermediate loop. The ability to produce a step change in torque with a step change in command generated from the outer loops represents the degree of control over the motor drive system for high-performance applications. Vector control in induction motors enables the machine to produce step changes in torque with instantaneous transition from one steady state to another steady state, which dramatically improves the dynamic performance of the drive system.

The objective of vector control or field orientation is to make the induction motor emulate the separately excited DC motor or the PM brushless DC motor. To understand vector control, let us revisit the torque production mechanism in DC machines. A simplified diagram of a DC motor with the field produced by separate excitation is shown in Figure 8.14. The field flux linkage space vector λ_f is stationary and is along the d-axis of the motor. The armature current space vector $\vec{i}_a$ is always along the q-axis due to action of the commutator and brushes, even though the rotor is revolving. The orthogonality between the field and armature current ensures the optimal condition for torque production, providing the highest torque-per-ampere ratio. The electromagnetic torque of the DC machine is given by

$$T_e = k_T \lambda_f i_a \qquad (8.16)$$

where k_T is a machine constant depending on construction and size of the machine. Vector notations are dropped, because these variables are constants in DC machines. The armature and field circuits in a separately excited DC machine are completely independent or decoupled, allowing independent control over torque and magnetic field. Independent flux control is especially desirable for EV-type applications, where flux weakening is used at higher speeds above rated torque conditions in the constant power region of torque–speed characteristics. The constant power range helps minimize the transmission gear requirements for propulsion drives.

In the case of PM brushless DC motors (i.e., PM trapezoidal machines), the rotor position sensor and power electronic converter replace the mechanical commutators and brushes of DC motors and work in synchronism to maintain the orthogonality between stator current space vector $\vec{i}_s(t)$ and the rotor flux vector $\vec{\lambda}_r(t)$ on an average basis. The back-emfs in these machines are trapezoidal and not

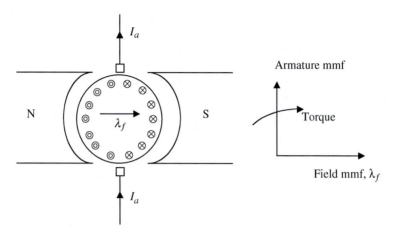

FIGURE 8.14 Torque in a separately excited DC machine.

sinusoidal; hence, vector control is not possible. Torque ripple is a dominating problem in PM brushless DC motors, because square-wave currents are used for torque production in order to synchronize with the trapezoidal back-emfs. In contrast, the orthogonality between armature and field mmfs is continuously maintained in DC commutator machines that help deliver smoother torque. However, in the case of PM sinusoidal machines, the back-emfs are sinusoidal, and using vector control, smooth torque control can be achieved like in an induction machine.

In light of the discussion presented above, the primary requirements of instantaneous torque control are controllability of the armature current, a controlled or constant field flux, and an orthogonal spatial angle between stator mmf axis and rotor mmf axis. In the case of DC and PM brushless DC motors, the last two requirements are automatically met with the help of commutators and brushes, and position sensors and inverter switching, respectively. However, bear in mind that orthogonality is maintained on an average basis only for PM brushless DC motors, the effect of which shows up in performance. In the case of induction machines and PM sinusoidal machines, these requirements are met with the help of dq-models and reference frame transformations. Instantaneous torque control is achieved when the three requirements are met at every instant of time. Note that the armature of a machine is the component that carries the bulk of the current delivered by the source. In the case of DC machines, the armature is in the rotor, while in the case of AC machines, the armature is in the stator. The control of armature currents is achieved with the help of current regulators, such as the hysteresis current regulator or a PI current regulator. Armature current control is necessary to overcome the effects of armature winding resistance, leakage inductance, and induced voltage in the stator windings. For field-weakening operation, the rotor flux needs to be reduced, which is achieved through field current control. The task is simple in separately excited DC machines. In induction and PM sinusoidal machines, dq modeling that decouples the torque and flux, producing components of currents and subsequent control of these components, help achieve the objective.

8.2.1 VECTOR CONTROL OF INDUCTION MOTORS

Vector control refers to the control of magnitude and phase angles of the stator and rotor critical components. The term *field orientation* is used as a special case of vector control representing a spatial angle of 90° between the rotor and stator critical components. The term vector control is general and is also used for controls where the angle between the rotor and stator critical components is something other than 90°.

Transformation of the variables into a rotating reference frame facilitates the instantaneous torque control of an induction machine, similar to that of a DC machine. Consider the electromagnetic torque expression (Equation 5.52) in the *dq* reference frame,

$$T_e = \frac{3}{2}\frac{P}{2}\frac{L_m}{L_r}\left(\lambda_{dr}i_{qs} - \lambda_{qr}i_{ds}\right)$$

If we choose a reference frame that rotates at synchronous speed with the rotor flux linkage vector $\bar{\lambda}_{qdr}(t)$ continuously locked along the *d*-axis of the *dq* reference frame, then $\lambda_{qr} = 0$. The resulting torque equation is then

$$T_e = \frac{3}{2}\frac{P}{2}\frac{L_m}{L_r}\lambda_{dr}i_{qs} \qquad (8.17)$$

Note the similarity between the above torque expression and the DC motor torque expression of Equation 8.16. An instantaneous change in i_{qs} current with constant λ_{dr} will result in an instantaneous change in torque similar to the situation in DC motors. We can conclude that an AC machine can be made to appear like a DC machine with appropriate reference frame transformations and, hence, can be controlled similarly. There are several ways of implementing the vector control on induction motors. Implementation can be carried out through transformation into any one of the several synchronously rotating reference frames, such as the rotor flux reference frame, air gap flux reference frame, and stator flux reference frame.[5,6] The task is the simplest in the rotor flux oriented reference frame, which will be described in the next section. In addition, according to the method of rotor flux angle measurement, the vector control is labeled as either direct vector control or indirect vector control.

8.2.2 ROTOR FLUX-ORIENTED VECTOR CONTROL

The rotor flux linkage vector direction is chosen as the *d*-axis of the reference frame in rotor flux-oriented vector control methods. The reference frame is also assumed to be rotating at the speed of the rotor flux vector. In the rotor flux-oriented reference frame, *q*-axis rotor flux $\lambda_{qr}^{rf} = 0$, and the torque is

$$T_e = \frac{3}{2}\frac{P}{2}\frac{L_m}{L_r}\lambda_{dr}^{rf}i_{qs}^{rf} \qquad (8.18)$$

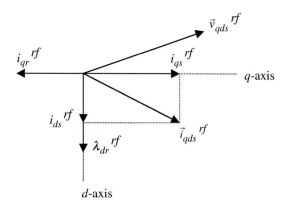

FIGURE 8.15 *d*- and *q*-axes currents in the rotor-flux-oriented reference frame.

The instantaneous control of the *q*-axis stator current in the rotor flux reference frame results in an instantaneous response of the motor torque, provided that the rotor flux is held constant. The current controller can be assumed to have a high bandwidth so that it establishes the stator command currents instantaneously. Therefore, the dynamics of the controller are primarily associated with the rotor circuit. The rotor flux and current and the stator current in the rotor flux oriented reference frame are shown in Figure 8.15. Rotor voltage and current equations in the rotor flux reference frame, derived from Equation 5.45, are

$$v_{qr}^{rf} = R_r i_{qr}^{rf} + p\lambda_{qr}^{rf} + \left(\omega_{rf} - \omega_r\right)\lambda_{dr}^{rf}$$

$$v_{dr}^{rf} = R_r i_{dr}^{rf} + p\lambda_{dr}^{rf} - \left(\omega_{rf} - \omega_r\right)\lambda_{qr}^{rf}$$

The flux linkages from Figure 5.28 are

$$\lambda_{qr}^{rf} = L_{lr} i_{qr}^{rf} + L_m \left(i_{qs}^{rf} + i_{qr}^{rf}\right) = L_m i_{qs}^{rf} + L_r i_{qr}^{rf}$$

$$\lambda_{dr}^{rf} = L_{lr} i_{dr}^{rf} + L_m \left(i_{ds}^{rf} + i_{dr}^{rf}\right) = L_m i_{ds}^{rf} + L_r i_{dr}^{rf}$$

The rotor voltages are identically zero in squirrel cage induction machines. Moreover, the *q*-axis flux linkage is zero in the rotor flux-oriented reference frame ($\lambda_{dr}^{rf} = 0$). Therefore, the rotor *dq*-circuit voltage balance equations become

$$0 = R_r i_{qr}^{rf} + \left(\omega_{rf} - \omega_r\right)\lambda_{dr}^{rf} \tag{8.19}$$

$$0 = R_r i_{dr}^{rf} + p\lambda_{dr}^{rf} \tag{8.20}$$

An important relation required to implement one form of vector control (indirect method) is the slip relation, which follows from Equation 8.19 as

$$\omega_{rf} - \omega_r = s\omega_{rf} = \frac{R_r i_{qr}^{rf}}{\lambda_{dr}^{rf}} = \frac{R_r}{L_r} \frac{L_m i_{qs}^{rf}}{\lambda_{dr}^{rf}} \tag{8.21}$$

From the q-circuit flux linkage expression, we get:

$$\lambda_{qr}^{rf} = L_m i_{qs}^{rf} + L_r i_{qr}^{rf} = 0 \quad \Rightarrow \quad i_{qr}^{rf} = -\frac{L_m}{L_r} i_{qs}^{rf} \tag{8.22}$$

Equation 8.22 describes that there will be an instantaneous response without any delay in the rotor current following a change in the torque command stator current i_{qs}^{rf}. The major dynamics in the rotor flux-oriented controller is in the behavior of λ_{dr}^{rf}, which is related to the d-axis rotor and stator currents according to

$$\lambda_{dr}^{rf} = L_r i_{dr}^{rf} + L_m i_{ds}^{rf} \tag{8.23}$$

Now, using Equations 8.20 and 8.23, the following dynamic relations can be easily established:

$$i_{dr}^{rf} = -\frac{L_m P}{R_r + L_r p} i_{ds}^{rf} \tag{8.24}$$

$$\lambda_{dr}^{rf} = \frac{R_r L_m}{R_r + L_r p} i_{ds}^{rf} \tag{8.25}$$

Equation 8.24 shows that i_{dr}^{rf} exists only when i_{ds}^{rf} changes and is zero in the steady state. Equation 8.25 gives the dynamics associated in bringing a change in the rotor flux λ_{dr}^{rf}. In steady state, the rotor flux is

$$\lambda_{dr}^{rf} = L_m i_{ds}^{rf}$$

Therefore, in order to change the rotor flux command, i_{ds}^{rf} must be changed, which will cause a transient occurrence of d-axis rotor current i_{dr}^{rf}. The time constant associated with these dynamics is $\tau_r = L_r/R_r$, which is commonly known as the rotor time constant.

8.2.3 DIRECT AND INDIRECT VECTOR CONTROL

The key to implementation of vector control is to find the instantaneous position of the rotor flux with respect to a stationary reference axis. Let this angle be defined

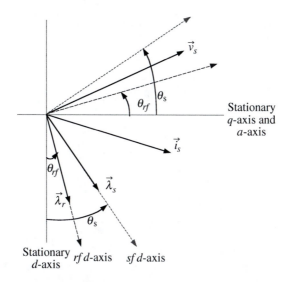

FIGURE 8.16 Relationship among stationary, rotor flux (*rf*) oriented and stator flux (*sf*) oriented reference frames.

as θ_{rf} with respect to the Phase-*a* reference axis, as shown in Figure 8.16. Vector control methods can be implemented in one of two ways, defined as direct and indirect methods according to the nature of measuring or calculating the rotor flux angle.

8.2.3.1 Direct Vector Control

In direct vector control methods, the rotor flux angle is calculated from direct measurements of machine electrical quantities. Using the *dq* model of the machine, the measurements are used to calculate the rotor flux vector, which directly gives the rotor flux angle θ_{rf}. The measurements of electrical variables can be carried out in one of several different ways. For example, flux-sensing coils or Hall sensors placed in the air gap can be used to measure the air gap flux λ_{qdm}. The subscript *m* stands for mutual flux in the air gap between the stator and the rotor. The *dq* model of the induction machine gives the following mathematical relations:

$$\vec{\lambda}^s_{qdm} = L_m \left(\vec{i}^s_{qds} + \vec{i}^s_{qdr} \right) \tag{8.26}$$

and

$$\vec{\lambda}^s_{qdr} = L_m \vec{i}^s_{qds} + L_r \vec{i}^s_{qdr} \tag{8.27}$$

The superscript *s* is used for the reference frame, because the air gap flux measurements are, with respect to the stator, in the stationary reference frame. Now, the objective is to write $\vec{\lambda}^s_{qdr}$ in terms of measurable quantities. From Equation 8.26, the rotor current can be written as

$$\vec{i}^s_{qdr} = \frac{\vec{\lambda}^s_{qdm}}{L_m} - \vec{i}^s_{qds}$$

Substituting the rotor current in Equation 8.27,

$$\vec{\lambda}^s_{qdr} = \frac{L_r}{L_m}\vec{\lambda}^s_{qdm} - \left(L_r - L_m\right)\vec{i}^s_{qds}$$

$$\Rightarrow \lambda_r = \left|\vec{\lambda}^s_{qdr}\right|, \quad \theta_{rf} = \angle\vec{\lambda}^s_{qdr}$$

The approach is straightforward, requiring knowledge of only two motor parameters, rotor leakage inductance $L_r - L_m = L_{lr}$ and the ratio L_r/L_m. The rotor leakage inductance is a fairly constant value, while the L_r/L_m ratio varies only a little by magnetic flux path saturation. However, the big disadvantage is the need for flux sensors in the air gap.

The flux sensors can be avoided by using voltage and current sensors to measure the stator applied voltages and currents. The measurements are used to calculate the stator flux by direct integration of the phase voltage:

$$\vec{\lambda}^s_{qds} = \int\left(\vec{v}^s_{qds} - R_s\vec{i}^r_{qds}\right)dt$$

The rotor flux vector can be obtained from the stator flux vector using the following mathematical relationship:

$$\vec{\lambda}^s_{qdr} = \frac{L_r}{L_m}\left(\vec{\lambda}^s_{qds} - L'_s\vec{i}_{qds}\right)$$

where

$$L'_s = L_s - \frac{L_m^2}{L_r}$$

The method requires the knowledge of three motor parameters R_s, L'_s, and L_r/L_m. The major difficulty of the method is significant variation of the stator resistance R_s due to temperature dependence and integration of the phase voltage to obtain stator

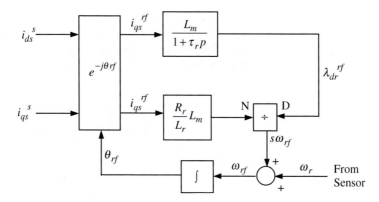

FIGURE 8.17 Rotor flux angle calculation in indirect vector control method.

flux at low speeds. Integration is especially inaccurate at low speeds due to the dominating resistive voltage-drop term.

8.2.3.2 Indirect Vector Control

In the indirect vector control scheme, which is more commonly used, a rotor position sensor is used to derive the speed and rotor position information, and the slip relation of Equation 8.22 is used to derive the rotor flux angle. The implementation of rotor flux angle calculation in the indirect vector control method is shown in Figure 8.17.

8.2.3.3 Vector Control Implementation

The vector-controlled drive has three major components like any other motor drive system: electric machine, power converter, and controller. An implementation block diagram of a speed-regulated vector-controlled drive is shown in Figure 8.18. The controller processes the input command signals and the feedback signals from the converter and the motor and generates the gating signals for the PWM inverter or converter.

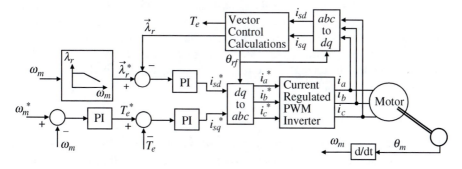

FIGURE 8.18 Implementation of vector control.

In the closed-loop speed-controlled system, the input is a speed reference, which is compared with the measured speed feedback signal to generate the control signals for maintaining the desired speed. The vector controller requires reference frame transformations and several computations that are typically implemented in a digital signal processor. The controller outputs in the first stage are the three-phase reference currents $i_a^*(t)$, $i_b^*(t)$, and $i_c^*(t)$ for the current regulated PWM inverter. The current controller in the second stage generates the PWM gating signals for the power electronics converter switches to establish the desired currents in the electric motor. Sensors provide current feedback information to the controller for vector calculations. Vector calculations involve transformation to a suitable reference and then torque and reference frame angle calculation. The torque and angle calculation can be in the rotor flux reference frame, where torque Equation 8.21 and the block diagram of Figure 8.17 are used. For speeds up to the rated speed of the machine, rated rotor flux $\tilde{\lambda}_r$ is used. For higher speeds, the flux command is reduced to operate the machine in the constant power mode. This mode of operation is known as the flux weakening mode. In the indirect control method, position and speed sensors are used that provide rotor position and motor speed feedback information (the implementation shown in Figure 8.17).

8.3 PM SYNCHRONOUS MOTOR DRIVES

The advantage of a PM synchronous machine is that it can be driven in the vector-controlled mode, delivering high performance, unlike the PM trapezoidal machine, which has to be driven more in the six-step mode. Of course, high-precision position information is needed to implement vector control in PM synchronous machines. A typical PM synchronous motor drive consists of a PM synchronous motor, a three-phase bridge inverter, gate drivers, position sensor, current or voltage sensors, a microprocessor, and its interfacing circuits, as shown in Figure 8.19.

Vector control of a PM synchronous motor is simpler than that of an induction motor, because the motor always rotates at synchronous speed. In vector calculations, only the synchronously rotating reference frame is necessary. The system controller sets the reference or command signal, which can be position, speed, current, or torque. The variables needed for the controller are the feedback signals from the sensing circuits (position, speed, current, or voltage) or estimated values in the signal processor. The error signals between the reference and actual variable signals are transformed to gate control signals for the inverter switches. The switches follow the gate commands to decrease the error signals by injecting desired stator currents into the three-phase stator windings.

8.3.1 VECTOR CONTROL

Vector control is used for the PM synchronous motor drive in EV and HEV applications to deliver required performance. Torque Equation 6.2 for the PM synchronous motor shows that if the d-axis current is maintained constant, the generated torque

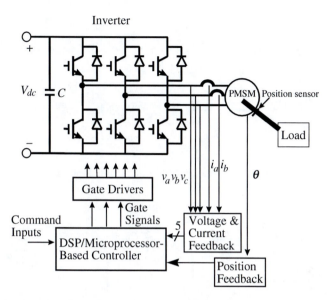

FIGURE 8.19 A typical PM synchronous motor drive structure.

is proportional to the q-axis current. For the special case when i_d is forced to be zero, $\lambda_d = \lambda_{af}$ and

$$T_e = \frac{3}{2} \cdot \frac{P}{2} \cdot \lambda_{af} \cdot i_q$$

$$= k_e i_q \tag{8.28}$$

where

$$k_e = \frac{3}{2} \cdot \frac{P}{2} \cdot \lambda_{af}$$

= motor constant. Because the magnetic flux linkage is a constant, torque is directly proportional to q-axis current. The torque equation is similar to that of a separately excited DC machine. Therefore, using reference frame transformations, the PM synchronous motor can be controlled like a DC machine.

8.3.2 FLUX WEAKENING

The PM synchronous motor can be operated in the field-weakening mode, similar to the DC motor, to extend the constant power range and achieve higher speeds. An injection of negative i_d will weaken the air gap flux as seen from Equation 6.2. The implementation technique for the field-weakening mode is shown in Figure 8.20.

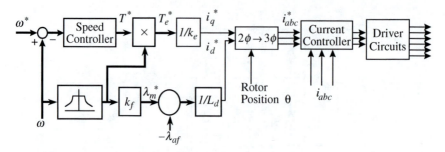

FIGURE 8.20 Vector controller structure for PMSM with field weakening.

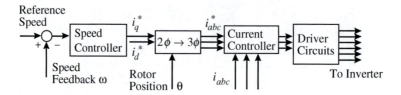

FIGURE 8.21 Current controller block diagram.

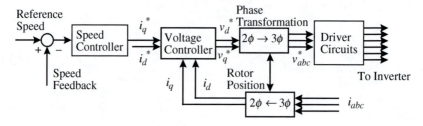

FIGURE 8.22 Voltage controller block diagram.

8.3.3 CURRENT AND VOLTAGE CONTROLLERS

Current and voltage control techniques, as described in Section 8.1, are applicable to the PM synchronous motor drive. PM synchronous motor drives can be designed using a current controller with a dual-phase mode. The simplified block diagram of a current controller is shown in Figure 8.21. The drawback of the current controller can be avoided if the current error signals drive a linear controller to convert the current commands into voltage commands. The voltage command signals can then be used for a voltage PWM scheme, such as the sinusoidal PWM or the space vector PWM. The block diagram of such a method is shown in Figure 8.22.

8.4 SR MOTOR DRIVES

The power electronic drive circuits for SR motor drives are different from those of AC motor drives. The torque developed in an SR motor is independent of the direction of current flow. Therefore, unipolar converters are sufficient to serve as the power converter circuit for the SR motor, unlike induction motors or synchronous motors, which require bidirectional currents. This unique feature of the SR motor, together with the fact that the stator phases are electrically isolated from one another, generated a wide variety of power circuit configurations. The type of converter required for a particular SR motor drive is intimately related to motor construction and the number of phases. The choice also depends on the specific application.

8.4.1 SRM CONVERTERS

The most flexible and the most versatile four-quadrant SRM converter is the bridge converter, shown in Figure 8.23a, which requires two switches and two diodes per phase.[7,8] The switches and the diodes must be rated to withstand the supply voltage plus any transient overload. During the magnetization period, both switches are turned on, and the energy is transferred from the source to the motor. Chopping or PWM, if necessary, can be accomplished by switching either or both of the switches during the conduction period, according to the control strategy. At commutation, both switches are turned off, and the motor phase is quickly defluxed through the freewheeling diodes. The main advantage of this converter is the independent control of each phase, which is particularly important when phase overlap is desired. The only disadvantage is the requirement of two switches and two diodes per phase. This converter is especially suitable for high-voltage, high-power drives.

The split-capacitor converter shown in Figure 8.23b has only one switch per phase but requires a split DC supply.[7] The phases are energized through the upper or the lower DC bus rail and the midpoint of the two capacitors. Therefore, only one-half the DC bus voltage can be applied for torque production. In order to maintain power flow balance between the two supply capacitors, the switching device and the freewheeling diode are transposed for each phase winding, which means that the motor must have an even number of phases. Also, the power devices must be rated to withstand the full DC supply voltage.

In low-speed applications, where PWM current control is desirable over the entire range of operation, the bridge converter can be reduced to the circuit shown in Figure 8.23c, developed by Miller.[8] In this converter, chopping is performed by one switch common to all phases. The circuit requires $(n + 1)$ switches for an n-phase motor. The main limitation of this circuit is that at higher speeds, the off-going phase cannot be de-energized fast enough, because the control switch Q_1 keeps turning on intermittently, disabling forced demagnetization. A class of power converter circuits with less than two switches per phase for SR motors having four or more phases has been developed by Pollock and Williams.[9]

The energy-efficient C-dump converter shown in Figure 8.23d is a regenerative converter topology with a reduced number of switches.[10] The topologies were derived from the C-dump converter proposed earlier by Miller.[8] The energy-efficient converter topologies eliminate all the disadvantages of the C-dump converter without sacrificing its attractive features, and they also provide additional advantages. The attractive features of the converters include a lower number of power devices, full regenerative capability, freewheeling in chopping or PWM mode, simple control

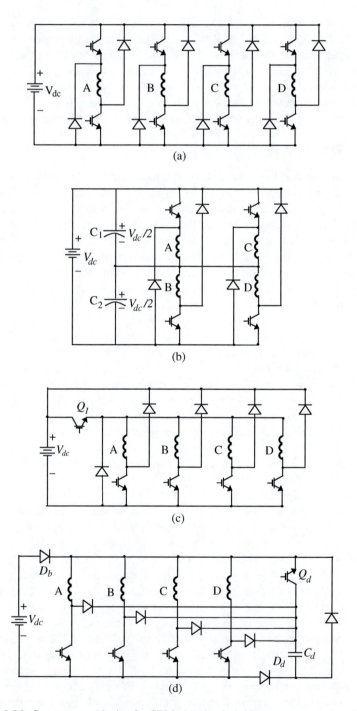

FIGURE 8.23 Converter topologies for SRM: (a) classic bridge power converter; (b) split-capacitor converter; (c) Miller converter; and (d) energy-efficient converter.

strategy, and faster demagnetization during commutation. The energy-efficient C-dump converter has one switch plus one diode forward voltage drop in the phase magnetization paths.

Converters with a reduced number of switches are typically less fault-tolerant compared to the bridge converter. The ability to survive component or motor phase failure should be a prime consideration for high-reliability applications. On the other hand, in low-voltage applications, the voltage drop in two switches can be a significant percentage of the total bus voltage, which may not be affordable. Among other factors to be considered in selecting a drive circuit are cost, complexity in control, number of passive components, number of floating drivers required, etc. The drive converter must be chosen to serve the particular needs of an application.

8.4.2 SRM CONTROLS

Appropriate positioning of the phase excitation pulses relative to the rotor position is the key in obtaining effective performance out of an SR motor drive system. The turn-on time, the total conduction period, and the magnitude of the phase current determine torque, efficiency, and other performance parameters. The type of control to be employed depends on the operating speed of the SRM.

8.4.2.1 Control Parameters

The control parameters for an SR motor drive are the turn-on angle (θ_{on}), turn-off angle (θ_{off}), and the phase current. The conduction angle is defined as $\theta_{dwell} = \theta_{off} - \theta_{on}$. The complexity of determination of the control parameters depends on the chosen control method for a particular application. The current command can be generated for one or more phases depending on the controller. In voltage-controlled drives, the current is indirectly regulated by controlling the phase voltage.

At low speeds, the current rises almost instantaneously after turn-on due to the negligible back-emf, and the current must be limited by controlling the average voltage or regulating the current level. The type of control used has a marked effect on the performance of the drive. As the speed increases, the back-emf increases, as explained before, and opposes the applied bus voltage. Phase advancing is necessary to establish the phase current at the onset of the rotor and stator pole overlap region. Voltage PWM or chopping control is used to force maximum current into the motor in order to maintain the desired torque level. Also, phase excitation is turned off early enough so that phase current decays completely to zero before the negative torque-producing region is reached.

At higher speeds, the SRM enters the single-pulse mode of operation, where the motor is controlled by advancing the turn-on angle and adjusting the conduction angle. At very high speeds, the back-emf will exceed the applied bus voltage once the current magnitude is high and the rotor position is appropriate, which causes the current to decrease after reaching a peak, even though a positive bus voltage is applied during the positive $dL/d\theta$. The control algorithm outputs θ_{dwell} and θ_{on} according to speed. At the end of θ_{dwell}, the phase switches are turned off so that negative voltage is applied across the phase to commutate the phase as quickly as possible. The back-emf reverses polarity beyond the aligned position and may cause the current to increase in this region, if the current does not decay to insignificant levels. Therefore, the phase commutation must precede the aligned position by several degrees so that the current decays before the negative $dL/d\theta$ region is reached.

In the high-speed range of operation, when the back-emf exceeds the DC bus voltage, the conduction window becomes too limited for current or voltage control, and all the chopping or PWM has to be disabled. In this range, θ_{dwell} and θ_{adv} are the only control parameters, and control is accomplished based on the assumption that approximately θ_{dwell} regulates torque and θ_{adv} determines efficiency.

8.4.2.2 Advance Angle Calculation

Ideally, the turn-on angle is advanced such that the reference current level i^* is reached just at the onset of pole overlap. In the unaligned position, phase inductance is almost constant, and hence, during turn-on, back-emf can be neglected. Also, assuming that the resistive drop is small, Equation 6.6 can be written as

$$V_{ph} = L(\theta)\frac{\Delta i}{\Delta \theta}\omega \tag{8.29}$$

Now, $\Delta i = i^*$ and $\Delta\theta = \theta_{overlap} - \theta_{on} = \theta_{adv}$, where $\theta_{overlap}$ is the position where pole overlap begins, θ_{on} is the turn-on angle, and θ_{adv} is the required phase turn-on advance angle. Therefore, we have

$$\theta_{adv} = L_u\omega\frac{i^*}{V_{dc}} \tag{8.30}$$

The above simple advance angle θ_{adv} calculation approach is sufficient for most applications, although it does not account for the errors due to neglecting the back-emf and the resistive drop in the calculation.

8.4.2.3 Voltage-Controlled Drive

In low-performance drives, where precise torque control is not a critical issue, fixed-frequency PWM voltage control with variable duty cycle provides the simplest means of control of the SRM drive. A highly efficient variable speed drive having a wide speed range can be achieved with this motor by optimum use of the simple voltage feeding mode with closed-loop position control only. The block diagram of the voltage-controlled drive is shown in Figure 8.24. The angle controller generates the turn-on and turn-off angles for a phase, depending on the rotor speed, which simultaneously determines the conduction period, θ_{dwell}. The duty cycle is adjusted according to the voltage command signal. The electronic commutator generates the gating signals based on the control inputs and the instantaneous rotor position. A speed feedback loop can be added on the outside, as shown when precision speed control is desired. The drive usually incorporates a current sensor typically placed on the

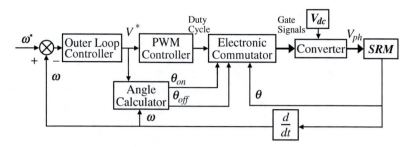

FIGURE 8.24 Voltage-controlled drive.

lower leg of the DC link for overcurrent protection. A current feedback loop can also be added that will further modulate the duty cycle and compound the torque–speed characteristics, just like the armature voltage control of a DC motor.

8.4.2.4 Current-Controlled Drive

In torque-controlled drives, such as in high-performance servo applications, the torque command is executed by regulating the current in the inner loop, as shown in Figure 8.25. The reference current i^* for a given operating point is determined from the load characteristics, the speed, and the control strategy. A wide-bandwidth current transducer provides the current feedback information to the controller from each of the motor phases. This mode of control allows rapid resetting of the current level and is used where fast motor response is desired. For loads with torque that increases monotonically with speed, such as in fans or blowers, speed feedback can be introduced in the outer loop for accurate speed control.

The simpler control strategy is to generate one current command to be used by all the phases in succession. The electronic commutator (see Figure 8.25) selects the appropriate phase for current regulation based on θ_{on}, θ_{off}, and the instantaneous rotor position. The current controller generates the gating signal for the phases based on the information coming from the electronic commutator. The current in the commuated phase is quickly brought down to zero, applying negative V_{dc}, while the incoming phase assumes the responsibility of torque production based on the commanded current. The phase transition in these drives is not smooth, which tends to increase the torque ripple of the drive.

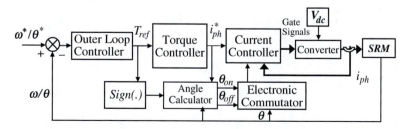

FIGURE 8.25 Current-controlled drive.

8.4.2.5　Advanced Control Strategies

A higher performance index, such as torque/ampere maximization, efficiency maximization, or torque ripple minimization can be required in certain applications. For example, in direct drive or traction applications, the efficiency over a wide speed range is critical. For applications, such as electric power steering in automobiles, the torque ripple is a critical issue. Typically, the torque/ampere maximization will go hand in hand with efficiency maximization, while torque ripple minimization will require the sacrifice of efficiency to a certain extent.

The high-performance drives will typically be current-controlled drives with sophistication added to the controller discussed earlier. For efficiency maximization, the key issue is the accurate determination of θ_{on} and θ_{off}, which may require modeling of the SRM and online parameter identification.[11] The modeling issue is equally important for torque ripple minimization, where the overlapping phase currents are carefully controlled during commutation.[12] In these sophisticated drives, the electronic commutator works in conjunction with the torque controller to generate the gating signals. The torque controller will include a model or tables describing the characteristics of the SRM.

The use of an indirect position sensing or sensorless method to eliminate the discrete position sensors is highly desirable for cost reduction and reliability enhancement purposes. All of the indirect position sensing methods developed for SR motors utilize the reluctance variation information along the air gap in one way or another. The position estimation is easily possible in SR motors, even at zero speed, because its inductance/flux varies in accordance with the rotor position. Some of the methods apply a diagnostic pulse in an unenergized phase to extract the rotor position information. A motor can be a good sensor of the motion, when its voltages and currents possess sufficient information to determine its position and velocity. Some of the observer-based methods depend on terminal measurements of voltages and currents.

REFERENCES

1. Trzynadlowski, A.M., *Introduction to Modern Power Electronics*, John Wiley & Sons, New York, 1998.
2. Dubey, G.K., *Power Semiconductor Controlled Drives*, Prentice Hall, New York, 1989.
3. Hou, C., DSP Implementation of Sensorless Vector Control for Induction Motors, MS thesis, University of Akron, OH, 2001.
4. Bose, B.K., *Modern Power Electronics and AC Drives*, Prentice Hall, New York, 2001.
5. Novotny, D.W. and Lipo, T.A., *Vector Control and Dynamics of AC Drives*, Oxford University Press, Oxford, 1996.
6. Mohan, N., *Electric Drives — An Integrated Approach*, MNPERE, Minneapolis, MN, 2001.
7. Davis, R.M., Ray, W.F., and Blake, R.J., Inverter drive for switched reluctance motor: circuits and component ratings, *IEE Proc.*, Vol. 128, B, No. 2, March, 1981, pp. 126–136.

8. Miller, T.J.E., *Switched Reluctance Motors and Their Control,* Magna Physics Publishing, Hillsboro, OH; Oxford Science Publications, Oxford, 1993.

9. Pollock, C. and Williams, B.W., Power converter circuits for switched reluctance motors with the minimum number of switches, *IEE Proc.,* Vol. 137, B, No. 6, November, 1990, pp. 373–384.

10. Mir, S., Husain, I., and Elbuluk, M., Energy-efficient C-dump converters for switched reluctance motors, *IEEE Transactions on Power Electronics,* Vol. 12, No. 5, September, 1997, pp. 912–921.

11. Mir, S., Husain, I., and Elbuluk, M., Switched reluctance motor modeling with online parameter adaptation, *IEEE Trans. on Industry Applications,* Vol. 34, No. 4, July–August, 1998, pp. 776–783.

12. Islam, M.S., Anwar, M.N., and Husain, I., A Sensorless Wide Speed Range SRM Drive with Optimally Designed Critical Rotor Angles, *IEEE-IAS Annual Conference Proc.,* Rome, 2000, pp. 1730–1737.

PROBLEMS

8.1

A 460 V, 60 Hz, six-pole, 1176 rpm, Y-connected induction motor has the following parameters referred to the stator at rated condition:

$$R_s = 0.19 \ \Omega, R_r = 0.07 \ \Omega, X_s = 0.75 \ \Omega, X_r = 0.67 \ \Omega, \text{ and } X_m = \infty.$$

The motor is fed by a six-step inverter. The inverter is fed from a battery pack through a DC/DC converter.

The battery pack voltage is 72 V. Neglecting all the losses:

(a) Determine the output of the DC/DC converter.
(b) Mention the type of the converter and its conversion ratio.

8.2

The motor in Problem 8.1 is employed to drive an EV that requires 300 N-m to propel the vehicle on a level road at constant velocity. The configuration is shown in Figure P8.2. Determine its operating speed and slip, while the frequency and voltage are kept constant at rated value.

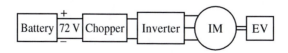

FIGURE P8.2

8.3

The vehicle in Problem 8.1 is moving downwards so that it requires 250 N-m.

(a) What will be the input voltage for the motor from the inverter? Hence, determine the conversion ratio of the converter. Frequency is kept constant at rated value, and the motor is running at the rated speed.
(b) What should be the operating frequency of the inverter if the input voltage to the motor is kept constant at rated value, and the motor is running at rated speed?

8.4

Find the speed of the motor mentioned in Problem 8.1 for a braking torque of 350 N-m and the inverter frequency of 40 Hz when the motor is supplied at rated voltage.

8.5

A three-phase induction machine is operated from a variable voltage, fixed frequency source.

(a) Derive an expression for machine efficiency in terms of slip (not in terms of torque and speed). Include only stator and rotor copper losses and core loss in P_{loss}. Model core loss by a constant resistance in the equivalent circuit. To simplify the analysis, assume that core loss resistance and magnetizing reactance are large compared to the other parameters. Under this assumption, you can use an approximate equivalent circuit, where the core loss resistance and magnetizing reactance are directly across the stator terminals.
(b) Does motor efficiency depend on terminal voltage? Calculate the slip that maximizes motor efficiency.

8.6

An AC inverter is operated in a sinusoidal pulse mode. The transistor base current waveforms are shown in Figure P8.6. Sketch line-to-line voltages v_{AB}, v_{BC}, and v_{CA}, and line to neutral voltage v_{AN} in the space provided. Briefly comment on the voltages. Are they balanced? (i_{ci} for i = 1 to 6 are the base currents for transistors 1 to 6, respectively).

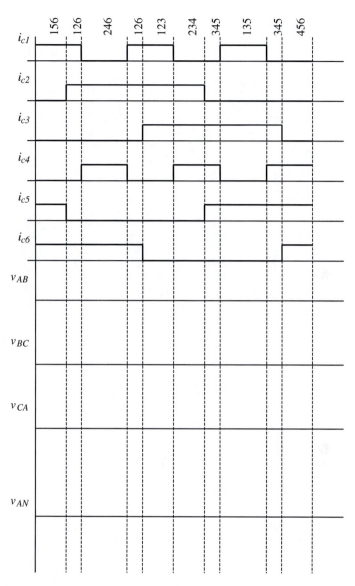

FIGURE P8.6

9 Electric Vehicle Drivetrain

The transmission elements and the propulsion unit combined are referred to as the drivetrain of the vehicle. The transmission is the mechanical linkage that transmits power between the electric motor shaft and the wheels. The drivetrain is also often referred to as the powertrain of the vehicle. The drivetrain of an electric vehicle (EV) consists of the electric motor, gearbox, driveshaft (only in rear-wheel drives), differential, half-shafts, and wheels. The ability of electric motors to start from zero speed and operate efficiently over a wide speed range makes it possible to eliminate the clutch that is used in internal combustion engine vehicles (ICEV). A single gear ratio is sufficient to match the wheel speed with the motor speed. EVs can be designed without a gear, but the use of a speed reducer allows the electric motor to operate at much higher speeds for given vehicle speeds, which minimizes the motor size because of the low torque requirement at higher speeds.

The transmission of an hybrid electric vehicle (HEV) is more complex than that of the EV because of the coupling necessary between the electric motor and the internal combustion (IC) engine. The HEV transmission will be discussed in the next chapter.

9.1 EV TRANSMISSION CONFIGURATIONS

In the case of front-wheel drive, the electric motor drives the gearbox, which is mounted on the front axle, as shown in Figure 9.1. This configuration is for an EV using a single propulsion motor. The single motor drives the transaxle on a common axis, delivering power to the two wheels differentially through a hollow motor shaft.[1]

The use of two motors driving two front wheels simplifies the transmission and eliminates the differential. Several configurations are possible with two propulsion motors driving two wheels. In one arrangement, the motors, mounted to the chassis, can be connected to the wheels through two short half-shafts. The suspension system of the vehicle isolates the wheels and its associated parts from the rest of the components of the vehicle for easier handling of the vehicle, depending on roadway conditions. The wheels are able to move freely without the weight of the motors when they are mounted on the chassis. In an alternate arrangement, the motors are mounted on the half-shafts with the motor driveshaft being part of the half-shaft. The half-shafts connect the wheels on one side and the chassis through a pivot on the other side. In-wheel mounting of motors is another arrangement possible in EVs. The difficulty in this case is that the unsprung weight of the vehicle increases due to motors inside the wheels, making traction control more complex. To minimize

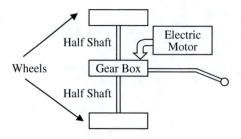

FIGURE 9.1 Typical front-wheel drive.

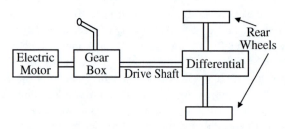

FIGURE 9.2 Typical rear-wheel drive.

the unsprung weight of the vehicle and because of the limited space available, the in-wheel motors must be of high-power density. As mentioned at the beginning, the use of a speed reducer is desirable, which adds to the constraint of limited space. The cost of a high-power, high-torque motor is the primary impediment in using in-wheel motors for EVs. Another problem with in-wheel motors is the heating due to braking compounded by the limited cooling capability in the restricted space. Nevertheless, the transmission simplicity has led to several projects for the development of in-wheel motors for EVs.

The transmission is more complex in the case of a rear-wheel drive, which requires a differential to accommodate unequal speeds of the inside and outside wheels of the rear axle during vehicle cornering. A typical rear-wheel drive transmission configuration is shown in Figure 9.2.

9.2 TRANSMISSION COMPONENTS

The gearbox (including the clutch or automatic transmission), driveshaft, and differential (in the case of a rear-wheel drive) are the major components of the transmission. The axles, wheels, and the braking systems are the auxiliary components of the transmission or powertrain. The output of the electric motor is the input to the transmission.

9.2.1 GEARS

The gear is a simple mechanical power transmission machine used to gain a mechanical advantage through an increase in torque or reduction in speed. This simple

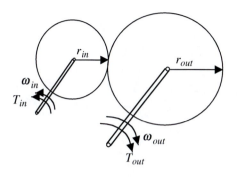

FIGURE 9.3

ω_{in} = Input angular speed
T_{in} = Input torque
ω_{out} = Output angular speed
T_{out} = Output torque
r_{in} = Input gear radius
r_{out} = Output gear radius

FIGURE 9.3 Gear mechanism.

mechanical device uses the law of conservation of energy, maintaining the steady flow of power or energy, because torque times speed is power that remains constant in the ideal transmission process. In an ideal gearbox, the motion is frictionless, and the power and energy supplied at the input point of the gearbox are equal to the power and energy available at the delivery point. The gearbox is not used to increase the shaft speed of an electric motor, because this means that a high-torque motor is unnecessarily designed, where the size of a motor is proportional to the torque output. Therefore, the gearbox can be used as a torque multiplier or speed reducer. A typical gear mechanism is shown in Figure 9.3.

Structurally, a gear is a round disk with teeth cut at equal intervals around the rim designed to engage with similar teeth of another disk. The round disks, placed in combination, transmit power from one gear to another. The teeth at both the disks lock the driving and the driven shafts together to transfer the energy through contact with little, if any, loss.

9.2.2 AUTOMOBILE DIFFERENTIAL

The automobile differential provides a mechanism of differential movement of the wheels on the rear axle. When a vehicle is turning a corner, the rear wheel to the outside of the curve must rotate faster than the inside wheel, because the former has a longer distance to travel. The type of gear used in an automobile differential is known as the planetary gear, where a set of gear trains operates in a coordinated manner. A simplified schematic of an automobile differential is given in Figure 9.4. The figure omits the teeth of the gears for simplicity. The driveshaft connected to the engine drives the pinion that is connected to the bevel (perpendicular) gear. The bevel gear is connected to the differential cage, which drives the wheel axles. The cage is connected to only one of the wheel axles, connecting the other axle only by means of the differential pinion. The differential pinion connecting the differential gears does not rotate as long as the speeds of the two wheels on the axle are the same. If one of the wheels slows due to cornering, the differential pinion starts rotating to produce a higher speed on the other wheel. The system described above is a simple differential that is not suitable for full torque transfer in low-traction conditions, such as on ice.[2]

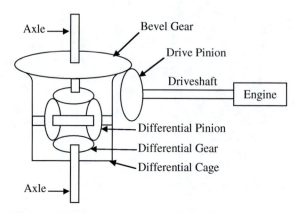

FIGURE 9.4 Simplified schematic of an automobile differential.

9.2.3 CLUTCH

The clutch is a mechanical device used to smoothly engage or disengage the power transmission between a prime mover and the load. The most common use of a clutch is in the transmission system of an automobile, where it links the IC engine with the rest of the transmission system of the vehicle. The clutch allows the power source to continue running, while the load is freely running due to inertia or is idle. For efficient operation of the IC engine, the wheel speed needs to be matched with the narrow range of high-torque operating speed of the IC engine. The clutch engages and disengages the IC engine from the road load as the gear ratio of the transmission is changed to match vehicle speed with the desired IC engine speed. Clutches can be eliminated in EVs, because the motor can start from zero speed and operate all the way to its maximum speed using a single gear ratio.

The clutch in the automotive transmission is an example of a positive clutch. There is a second general type of clutch, which is known as friction clutches, used to bring the rotational speed of one disk to that of the other disk. The clutching action can be brought about by electrical, pneumatic, or hydraulic action in addition to mechanical means discussed here.

9.2.4 BRAKES

The brakes in automobiles are mechanical clutches known as friction clutches, which use friction to slow a rotational disk. The driver controls the brake action through a foot-operated linkage. The friction clutch is composed of two disks, each connected to its own shaft. As long as the disks are not engaged, one disk can spin freely without affecting the other. When the rotating and the stationary disks are engaged through the operation action, friction between the two disks reduces the speed of the rotating disk. The kinetic energy of the vehicle transfers directly between the disks and is wasted due to friction.

The two types of brakes used in automobiles are disk and drum. Disk brakes have friction pads controlled by a caliper arrangement, which when engaged, clamps to a disk rotating with the wheel. The brake pads are designed to assist cooling and

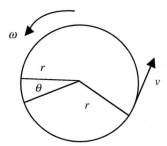

FIGURE 9.5 Variables in a gear.

resist fading. Fading causes the friction coefficient of braking to change with temperature rise. The high force required for caliper actuation is typically supplied from a power-assist device following the brake command input from the driver. Currently, hydraulic actuator systems are used for power assist, but the technology trend is to replace the hydraulic actuator system with an electric-motor-driven actuator system for superior performance.

Drum-type brake units have cylindrical surfaces and shoes instead of pads that hold the friction material. The shoes press against the drum cylinder upon a brake command input from the driver. The shoes can be arranged to press against the outer or inner surface of the rotating drum to retard wheel rotation. If the shoes are applied to the inner surface, then the centrifugal force of the drum due to rotation will resist disk engagement. If the shoes are pressed against the outer drum surface, the centrifugal force will assist engagement, but at the same time may cause overheating.

9.3 IDEAL GEARBOX: STEADY STATE MODEL

The EV transmission equations will be established assuming ideal gearbox assumptions, which are as follows:

1. $P_{losses} = 0 \Rightarrow$ Efficiency = 100%
2. Perfectly rigid gears
3. No gear backlash (i.e., no space between teeth)

The variables used in deriving the steady state model are given in Figure 9.3. Additional variables are shown in Figure 9.5, with respect to a single disk.

9.3.1 GEAR RATIO (GR)

For a disk with radius r, the tangential and the angular velocity are related by

$$\omega r = v$$

The tangential velocity at the gear teeth contact point is the same for the two gears disks shown in Figure 9.3 with different radii:

$$r_{in}\omega_{in} = v = r_{out}\,\omega_{out}$$

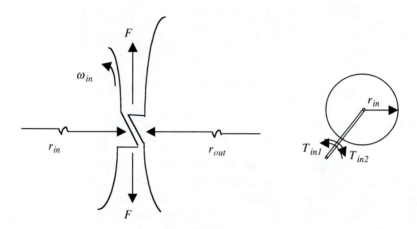

FIGURE 9.6 Force and torque working in a gear.

The gear ratio is defined in terms of the ratio of speed transformation between the input shaft and the output shaft.

$$GR = \frac{\omega_{in}}{\omega_{out}} = \frac{r_{out}}{r_{in}} \tag{9.1}$$

Assuming 100% efficiency of the gear train:

$$P_{out} = P_{in}$$

$$\Rightarrow T_{out}\omega_{out} = T_{in}\omega_{in}$$

The gear ratio in terms of the torque at the two shafts is

$$GR = \frac{T_{out}}{T_{in}} = \frac{\omega_{in}}{\omega_{out}} \tag{9.2}$$

The gear ratio can be alternately derived with the help of Figure 9.6. At the point of gear mesh, the supplied and delivered forces are the same. This is an example of Newton's third law of motion, which states that every action has an equal and opposite reaction. The torque at the shaft is the force at the mesh divided by the radius of the disk. In the two-gear combination, the torque ratio between the two gears is proportional to the ratio of gear disk radii.

The torque of the inner disk in terms of its radius and force at the gear mesh is

$$T_{in} = Fr_{in}$$

$$\Rightarrow F = \frac{T_{in}}{r_{in}}$$

Similarly, for the other disk with radius r_{out}, the force at the gear mesh is

$$F = \frac{T_{out}}{r_{out}}$$

Therefore, the gear ratio is

$$GR = \frac{T_{out}}{T_{in}} = \frac{r_{out}}{r_{in}} \tag{9.3}$$

9.3.2 Torque-Speed Characteristics

The electric motor is typically designed to operate at higher speeds to minimize the size of the motor. The gearbox functions as a torque multiplier to deliver high torque at a reduced speed at the vehicle wheels. The advantage of using a gear will be shown through a DC motor-driven EV system. Let the overall gear ratio between the electric motor and the vehicle wheel be GR with ω_m, and let T_m represent the motor speed and torque, respectively. The speed and torque at the wheels are ω_{out} and T_{out}, respectively. Part of the drivetrain for the EV is shown in Figure 9.7. For a separately excited DC motor, the speed-torque relationship at steady state is

$$\omega_m = \frac{V_t}{k\phi} - \frac{R_a}{(k\phi)^2} T_m \tag{9.4}$$

However,

$$\frac{\omega_m}{\omega_{out}} = GR = \frac{T_{out}}{T_m}$$

Substituting in Equation 9.4,

$$\omega_{out} = \frac{V_t}{GR(k\phi)} - \frac{R_a}{(GRk\phi)^2} T_{out} \tag{9.5}$$

Example 9.1

For the EV drivetrain shown in Figure 9.8, $V_t = 50$ V, $R_a = 0.7$ ohm, $k\phi = 1.3$, and $mg = 7848$N.

Gearbox gears are as follows: first $GR_1 = 2$, second $GR_2 = 1$, and $r_{wh} =$ wheel radius = 7 in = 0.178 m.

Find the vehicle maximum % gradability for each gear.

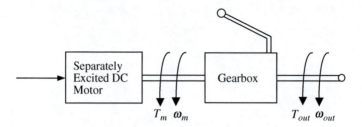

FIGURE 9.7 Connection of a gear with the motor.

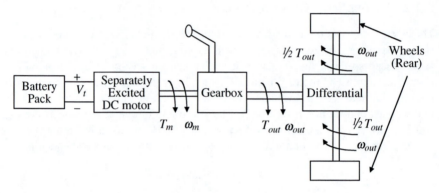

FIGURE 9.8 EV drivetrain for Example 9.1.

Solution

$$\text{Max. \% grade} = \frac{100\,F_{TR}}{\sqrt{(mg)^2 - F_{TR}^2}}$$

The torque–speed relationship is

$$\omega_{out} = \frac{V_t}{GR(k\phi)} - \frac{R_a}{(GRk\phi)^2}\,T_{out}$$

To find F_{TR}, set $\omega_{out} = 0$.

$$0 = \frac{V_t}{GR(k\phi)} - \frac{R_a}{(GRk\phi)^2}\,T_{out}$$

$$\Rightarrow T_{out} = \frac{V_t}{GR(k\phi)} * \frac{(GRk\phi)^2}{R_a} = \frac{GR(k\phi)V_t}{R_a}$$

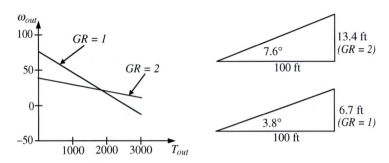

FIGURE 9.9 Plots for Example 9.1.

The relationship between F_{TR} and T_{out} is

$$F_{TR} = 2 \ (F_{TR} \text{ per rear wheel})$$

F_{TR} per rear wheel = torque per wheel/r_{wh} = (1/2 T_{out})/r_{wh}

Therefore,

$$F_{TR} = 2 \frac{\frac{1}{2} T_{out}}{r_{wh}} = \frac{T_{out}}{r_{wh}}$$

Substituting,

$$F_{TR} = \frac{(GRk\phi)V_t}{r_{wh}R_a} = \frac{GR(1.3)50}{(0.178)(0.7)} = 521.7 \ GR$$

In first gear, $F_{TR} = 1043$ N.

$$\text{Max. grade} = \frac{100(1043)}{\sqrt{(7878)^2 - (1043)^2}} = 13.4\%$$

In second gear, $F_{TR} = 521.7$ N, and maximum grade = 6.7%.

The steady state torque–speed characteristics for the separately excited DC motor connected to the gearbox for the first two gears are shown in Figure 9.9. The maximum percentage grades for the two gears are also shown in the figure.

9.4 EV MOTOR SIZING

Electric motors have three major segments in its torque–speed characteristics: constant torque region, constant power regions, and natural mode region. The envelope

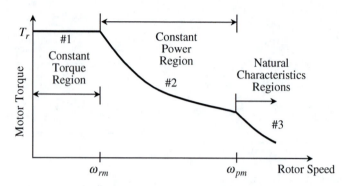

FIGURE 9.10 Electric motor torque–speed envelope.

of the electric motor torque–speed characteristics is shown in Figure 9.10. The motor delivers rated torque up to the base speed or rated speed ω_{pm} of the motor when it reaches its rated power condition. The motor rated speed is defined as the speed at which the motor can deliver rated torque at rated power. The motor operates in a constant power mode beyond the rated speed, where torque falls steadily at a rate that is inversely proportional to speed. Electric motors can operate at speeds higher than rated using field weakening in the constant power region. There is a third natural mode region for high motor speeds, where the torque falls rapidly, being inversely proportional to the square of the speed. The natural characteristic region can be an important part of the overall torque–speed curve of certain motors that can be used to reduce the power rating of the motor. However, in most cases, the vehicle's maximum speed is considered to be at the end of the constant power region. Note that the curves in Figure 9.10 show the envelope, i.e., the operating torque and speed limits in different regions. The electric motor can operate at any point within the envelope through the feed from a power electronics-based motor drive component. The salient feature of wide-operating speed range characteristics of an electric motor makes it possible to eliminate multiple gear ratios and the clutch in EV and other applications. A single gear ratio transmission is sufficient for linking the electric motor with the driveshaft. Electric motors with extended constant power region characteristics are what is needed to minimize the gear size in EVs.

The size of an electric motor depends on the maximum torque required from the machine. The higher the maximum torque required, the larger will be the size of the motor. In order to minimize the size and weight, electric motors are designed for high-speed operation for a given power rating. Gears are used to match the higher speed of the electric motor with the lower speed of the wheels. Typical motor speeds can be in the vicinity of 15,000 rev/m for typical wheel speeds of around 1000 rev/m for lightweight passenger vehicles. The transmission gear achieves this speed reduction in the range of ~10 to 15:1, typically in two stages of 3 to 4:1 of speed reduction. The gear sizing depends on whether the low speed or the high speed performance of the EV is more important based on the power rating determined for the EV.

The tractive force vs. speed characteristics of the propulsion system can be widely different for two gear ratios, as shown in Figure 9.11. Note that the rated speeds shown are for the drivetrain unit comprising the electric motor and transmission

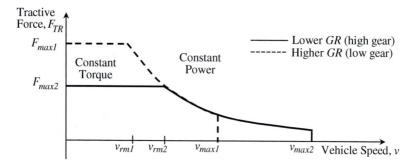

FIGURE 9.11 Electric motor torque–speed characteristics in terms of traction force and vehicle speed for two gear ratios.

system, and that the electric motor rated speed is different from these values. The electric motor speed can first be converted into drivetrain unit speed or vehicle wheel speed using the gear ratio as shown earlier. The motor rated speed at the wheel $\omega_{rm,wh}$ in rad/s can be converted to linear vehicle speed using $v_{rm} = \omega_{rm,wh} \cdot r_{wh}$, where r_{wh} is the radius of the wheel. The symbol v_{rm} is referred to here as the equivalent vehicle linear speed after accounting for the transmission gear and the wheel radius. A higher wheel speed or vehicle speed can be attained with a smaller gear ratio, but the peak traction force that the drivetrain can deliver will be smaller. The smaller traction force will limit the initial acceleration and maximum gradability capabilities of the vehicle. On the other hand, if a high gear ratio is used in the transmission for the same electric motor, the peak tractive force available at the wheels will be higher, but the maximum vehicle speed v_{max} will be limited. Therefore, we can conclude that the gear ratio depends on the rated motor speed, vehicle rated speed, vehicle maximum speed, wheel radius, and maximum gradability. It must be observed that a higher gear ratio entails a larger gear size. Therefore, the gear ratio and the electric motor rated speed must be selected simultaneously to optimize the overall size and performance requirements.

As outlined in Chapter 2, the complete design of the powertrain is a complex issue involving numerous variables, such as rated motor power P_m, rated motor speed, ω_{rm}, rated wheel speed, ω_{fwh}, rated vehicle speed v_f, etc., and numerous system parameters, such as vehicle total mass, rolling resistance coefficient, aerodynamic drag coefficient, etc. The design process starts with a set of known parameters and some educated estimate and ends with final design values that meet the requirements after several iterations. In the following, the design of the electric motor will be discussed in view of the specified requirements of the initial acceleration, rated velocity on a given slope, maximum steady state velocity, and maximum gradability.

9.4.1 INITIAL ACCELERATION

The initial acceleration is specified as 0 to v_f in t_f s. v_f is the vehicle rated speed obtained from $v_f = \omega_{fwh} \cdot r_{wh}$. The design problem is to solve for F_{TR} starting with a set of parameters including vehicle mass, rolling resistance, aerodynamic drag coefficient, percent grade, wheel radius, etc., some of which are known, while others have to be assumed. The acceleration of the vehicle in terms of these variables is given by Equation 2.24, repeated here for convenience:

$$a = \frac{dv}{dt} = \frac{F_{TR} - F_{RL}}{m}$$

The motor power rating can be obtained by solving the above differential equation for a given force-velocity profile, such as one of the two shown in Figure 9.11, and the following boundary conditions:

At $t = 0$, vehicle velocity $v = 0$.
At $t = t_f$, vehicle velocity $v = v_f$.

Integrating the differential equation within the interval $t = 0$ to $t = t_f$ for velocities 0 to v_f,

$$m \int_0^{v_f} \frac{dv}{F_{TR} - F_{RL}(v)} = \int_0^{t_f} dt$$

The rated vehicle velocity is higher than the rated motor velocity and lies in the constant power region of motor torque–speed characteristics. Splitting the integral on the left side into two velocity regions of 0-v_{rm} for the constant torque mode and of v_{rm}-v_f for the constant power mode, one can write:

$$m \int_0^{v_{rm}} \frac{dv}{\dfrac{P_m}{v_{rm}} - F_{RL}(v)} + m \int_{v_{rm}}^{v_f} \frac{dv}{\dfrac{P_m}{v} - F_{RL}(v)} = t_f \tag{9.6}$$

The road load resistance force F_{RL} can be expressed as a function of velocity, as shown in Chapter 2 for given values of rolling resistance, aerodynamic drag force, and roadway slope. Equation 9.6 can then be solved for motor power rating P_m for specified vehicle-rated velocity v_f and rated motor speed. Note that Equation 9.6 is a transcendental equation with F_{RL} being a function of velocity and can be solved numerically to find the motor power rating P_m. In fact, extensive computer computation and simulation aids a practical design to derive the required motor power rating and gear ratio of the powertrain.

An interesting analysis has been presented in the literature[3] to stress the importance of extended constant power region of motor torque–speed characteristics. It has been shown in the literature[3] that for $F_{RL} = 0$, the motor power rating is

$$P_m = \frac{m}{2t_f}\left(v_{rm}^2 + v_f^2\right) \tag{9.7}$$

Equation 9.7 shows that the motor power rating will be minimum when $v_{rm} = 0$, which means that the electric motor that operates entirely in the constant power mode is the smallest motor to satisfy the requirements. In the other extreme case, motor power will be double that of the smallest case, if the motor operates entirely in the constant torque mode with $v_{rm} = v_f$. Of course, eliminating the constant torque region and operating entirely in the constant power region are not practically realizable. In a practical setting, the electric motor should be designed with a low base speed or rated speed and a wide constant power region.

9.4.2 RATED VEHICLE VELOCITY

The drivetrain designed to accelerate the vehicle from zero to rated velocity will always have sufficient power to cruise the vehicle at rated speed, provided the roadway slope specified for initial acceleration has not been raised for rated velocity cruising conditions.

9.4.3 MAXIMUM VELOCITY

The tractive power required to cruise the vehicle at maximum vehicle velocity v_{max} is

$$P_{TR,\max} = mgv_{\max}\sin\beta + \left[mgC_1 + \frac{P}{2}A_F C_D \right]v_{\max}^3 + mgv_{\max}C_0 \qquad (9.8)$$

The dominant resistance force at high speeds is the aerodynamic drag force, with the power requirement to overcome it increasing at a cubic rate. For vehicles designed with fast acceleration characteristics, P_m is likely to be greater than $P_{TR,\max}$. If $P_{TR,\max} > P_m$ derived earlier to meet the initial acceleration requirement, then $P_{TR,max}$ will define the electric motor power rating. The natural mode region of electric motors can be used to meet very high maximum vehicle velocity requirements to minimize the motor size.

9.4.4 MAXIMUM GRADABILITY

The maximum gradability of a vehicle for a given motor and gear ratio can be derived from

$$\text{Max. \% grade} = \frac{100\,F_{TR}}{\sqrt{(mg)^2 - F_{TR}^2}}$$

The maximum tractive force F_{TR} available from the preliminary motor design can be plugged into the above equation to check whether the vehicle maximum gradability conditions are met or not. If the maximum electric motor power derived for acceleration or maximum vehicle velocity is not enough to meet the maximum gradability requirement of the vehicle, then either the motor power rating or the gear ratio has to be increased. Care must be taken not to violate the maximum vehicle velocity requirement when increasing the gear ratio. The gear ratio and motor power are decided in a coordinated manner to meet both requirements, while maintaining a reasonable size for both the electric motor and the gear.

REFERENCES

1. Willis, R.L. and Brandes, J., Ford next generation electric vehicle powertrain, *12th Electric Vehicle Symp.*, December, 1994, pp. 449–458.
2. Scott, T.E., *Power Transmission Mechanical, Hydraulic, Pneumatic, and Electrical*, Prentice Hall, New York, 2000.
3. Ehsani, M., Rahman, K.M., and Toliyat, H.A., Propulsion system design for electric and hybrid vehicles, *IEEE Transactions on Industrial Electronics*, Vol. 44, No. 1, February, 1997, pp. 19–27.

PROBLEM

9.1

An EV drivetrain employs a separately excited DC motor that drives the EV rear axle through a gearbox, as shown below in Figure P9.1. The vehicle is traveling in fourth gear on a level road at a constant velocity of 60 mi/h. All necessary parameters are as follows:

$$mg = 6867 \text{ N}, \ C_0 = 0.009, \ C_1 = 1.75 \text{ E-6 s}^2/\text{m}^2, \ A_F = 2 \text{ m}^2, \ C_d = 0.2,$$
$$\rho = 1.16 \text{ kg/m}^3, \ r_{wh} = \text{wheel radius} = 7.5 \text{ in, and gear ratio } GR = 0.4.$$

Motor parameters are as follows:

$$R_a = 0.2 \ \Omega, \ R_F = 150 \ \Omega, \ K = 0.8 \text{ V-s/Wb}, \ \phi = 3.75 \ I_F, \text{ and } B_g = 1.2 \ I_F.$$

In the following calculations, assume no power loss from motor output to wheels. Also, assume that chopper outputs are pure DC.

(a) Calculate the operating speed and torque of the motor.
(b) For $0.5 \leq I_F \leq 4$ A, plot I_A vs. I_F

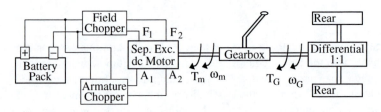

FIGURE P9.1

10 Hybrid Electric Vehicles

A hybrid electric vehicle (HEV) is a vehicle in which at least one of the energy sources, stores, or converters can deliver electric energy. A hybrid road vehicle is one in which the propulsion energy during specified operational missions is available from two or more kinds or types of energy stores, sources, or converters, of which at least one store or converter must be on board. The second definition of hybrid road vehicle is proposed by Technical Committee 69 of Electric Road Vehicles of the International Electrotechnical Commission.

The HEV serves as a compromise for the environmental pollution problem and the limited range capability of today's purely electric vehicle. HEVs have an electric motor as well as an internal combustion engine (ICE) to provide extended range and to "curve down" the pollution problem. Vehicle design complexity increases significantly with hybrid vehicles, because controls and support systems are needed for a thermal engine and an electric machine in addition to the components needed for controlled blending of power coming from the two sources. The hybrids are looked upon by many as a short-term solution until the range limitation and infrastructure problems of purely electric vehicles are solved. Nevertheless, a number of automotive manufacturers are marketing hybrid vehicles for the general population, with many others following suit.

This chapter focuses on familiarizing the reader with the basic drivetrains of HEVs. The basic thermodynamics of some of the heat engines or ICEs are discussed prior to discussing the design issues with HEVs.

10.1 TYPES OF HYBRIDS

10.1.1 SERIES AND PARALLEL HEVS

HEVs evolved out of two basic configurations: series and parallel. A series hybrid is one in which only one energy converter can provide propulsion power. The heat engine or ICE acts as a prime mover in this configuration to drive an electric generator that delivers power to the battery or energy storage link and the propulsion motor. The component arrangement of a series HEV is shown in Figure 10.1.

A parallel hybrid is one in which more than one energy source can provide propulsion power. The heat engine and the electric motor are configured in parallel, with a mechanical coupling that blends the torque coming from the two sources. The component arrangements of a parallel hybrid are shown in Figure 10.2.

Series HEV is the simpler type, where only the electric motor provides all the propulsion power. A downsized heat engine on board drives a generator, which supplements the batteries and can charge them when they fall below a certain state

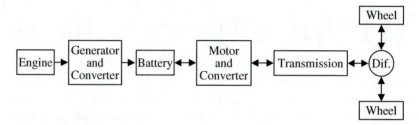

FIGURE 10.1 Series HEV drivetrain.

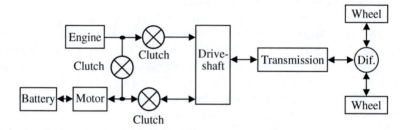

FIGURE 10.2 Parallel HEV drivetrain.

of charge. The power required to move the vehicle is provided solely by the electric motor. Beyond the heat engine and the generator, the propulsion system is the same as in an EV, making electric motor power requirements the same as for in the EV.

In parallel HEV, the heat engine and the electric motor are connected to the driveshaft through separate clutches. Power requirements of the electric motor in the parallel hybrid are lower than that of an EV or series hybrid, because the heat engine complements for the total power requirement of the vehicle. The propulsion power may be supplied by the heat engine, by the battery–motor set, or by the two systems in combination.

Series and parallel hybrids come in a variety of types. The mission of the vehicle and the optimum design for that mission dictate the choice. If the HEV is to be basically an EV with an ICE-assist for achieving acceptable range, then the choice should be a series hybrid, with the ICE ensuring that the batteries remain charged all the time. On the other hand, if the HEV is to be basically a vehicle with almost all the performance characteristics and comforts of an ICEV but with lower emission and fuel usage standards, then the choice should be a parallel configuration. Parallel HEVs have been built with performance that is equal, in all aspects of normal operation, to that of a conventional car. However, some series HEVs have also been built that perform nearly as well as ICEVs.

10.1.1.1 Advantages and Disadvantages

The advantages and disadvantages of series and parallel hybrids are summarized in the following. The advantages of a series HEV are:

1. Flexibility of location of engine-generator set
2. Simplicity of drivetrain
3. Suitability for short trips

The disadvantages of a series HEV are:

1. It needs three propulsion components: ICE, generator, and motor.
2. The motor must be designed for the maximum sustained power that the vehicle may require, such as when climbing a high grade. However, the vehicle operates below the maximum power most of the time.
3. All three drivetrain components need to be sized for maximum power for long-distance, sustained, high-speed driving. This is required, because the batteries will exhaust fairly quickly, leaving ICE to supply all the power through the generator.

The following are advantages of a parallel HEV:

1. It needs only two propulsion components: ICE and motor/generator. In parallel HEV, the motor can be used as the generator and vice versa.
2. A smaller engine and a smaller motor can be used to get the same performance, until batteries are depleted. For short-trip missions, both can be rated at half the maximum power to provide the total power, assuming that the batteries are never depleted. For long-distance trips, the engine may be rated for the maximum power, while the motor/generator may still be rated to half the maximum power or even smaller.

The following are disadvantages of a parallel HEV:

1. The control complexity increases significantly, because power flow has to be regulated and blended from two parallel sources.
2. The power blending from the ICE and the motor necessitates a complex mechanical device.

10.1.2 SERIES–PARALLEL COMBINATION

Although HEVs initially evolved as series or parallel, manufacturers later realized the advantages of a combination of the series and parallel configurations for practical road vehicles. In these combination hybrids, the heat engine is also used to charge the battery. The recently available Toyota Prius is an example of such a hybrid, where a small series element is added to the primarily parallel HEV. The small series element ensures that the battery remains charged in prolonged wait periods, such as at traffic lights or in a traffic jam. These combination hybrids can be categorically classified under parallel hybrids, because they retain the parallel structure of a component arrangement. It is important to stress the fact that the detailed configuration of an HEV depends on the application and the trade-off between cost and performance.

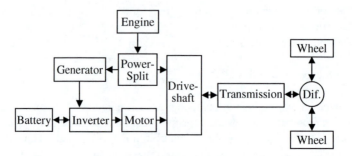

FIGURE 10.3 Series–parallel combination HEV.

The component arrangement of a series–parallel combination hybrid is shown in Figure 10.3. The schematic is based on the Toyota Prius hybrid design. A power split device allocates power from the ICE to the front wheels through the driveshaft and the electric generator, depending on the driving condition. The power through the generator is used to charge the batteries. The electric motor can also deliver power to the front wheels in parallel to the ICE. The inverter is bidirectional and is used to charge the batteries from the generator or to condition the power for the electric motor. For short bursts of speed, power is delivered to the driveshaft from the ICE and the electric motor. A central control unit regulates the power flow for the system using multiple feedback signals from the various sensors.

Use of the ICE to charge the batteries should be minimized when maximizing efficiency. Energy is always lost while charging and discharging the battery and during the power flow through the inverter. The vehicle should be operated off its engine or battery or both, until the battery is at a minimum acceptable state of charge, say 20 to 40%. The battery should be charged from the power grid when convenient.

10.2   INTERNAL COMBUSTION ENGINES

The devices that convert heat transfer to work cyclically are known as heat engines. Each cycle in the heat engine consists of several different *processes* or *strokes* (such as constant volume, constant pressure, and constant entropy processes) to convert thermal energy into useful work. Different types of cycles are available to design practical heat engines. The practical cycles have evolved due to the limitations of an ideal thermodynamics cycle (such as the Carnot cycle) that operates between the same heat addition and rejection temperatures, and also because of the difference in characteristics of the choices available for the energy source, working fluid, and hardware materials. It must be mentioned that although the heat engines described in the following undergo *mechanical cycles*, the working fluids do not execute a *thermodynamic cycle* because matter is introduced with one composition and is later discharged at a different composition.[1,2]

The heat engines of interest for EV and HEV applications, primarily the ICE and the gas turbine, will be discussed in this section. An ICE is a heat engine that utilizes fuel as a working fluid. The ICE uses heat cycles that gain its energy from the combustion of fuel within the engine. ICEs can be reciprocating type, where the reciprocating motion of a piston is converted to rotary motion through a crank mechanism. ICEs used in automobiles, trucks, and buses are of the reciprocating type, where the processes occur within a reciprocating piston–cylinder arrangements. The gas turbines used in power plants are also ICEs, where the processes occur in an interconnected series of different components. The Brayton cycle gas turbine engine has been adapted to the automotive propulsion engine and has the advantage of burning fuel that requires little refining and fuel that burns completely. Gas turbines have fewer moving parts, because there is no need to convert the rotary motion of the turbine. The disadvantages of gas turbines are complex construction and lower efficiency. Nevertheless, gas turbines are being considered for HEVs, and prototype vehicles have been developed.

The performance of a heat engine is measured by the efficiency of the heat engine cycle, defined as the ratio of the net work output per cycle W_{net} to the heat transfer into the engine per cycle. Another way of defining the performance of heat engines is to use the *mean effective pressure* P_{me}. The P_{me} is the theoretical constant gage pressure that, if exerted on the piston during the expansion stroke between the largest specific volume and the smallest volume, would produce the same net work as actually produced by the heat engine. Mathematically stated,

$$P_{me} = \frac{|W_{net}|}{displacement\ volume}$$

The heat engine cycle performance analysis is carried out from the information available at certain convenient state points in the cycle. The parameters needed at the state points are pressure, temperature, volume, and entropy. If two parameters are known at two state points, then the unknown parameters are usually obtained from the process that the working fluid undergoes between the two state points (such as constant pressure, isentropic, etc.) and the laws of thermodynamics. Discussions of the laws of thermodynamics and efficiency analysis of heat engine cycles are beyond the scope of this book. Only a general introduction to the heat engine cycles of interest will be given.

Prior to continuing the discussion on ICEs, a few words about entropy are in order. Entropy is a property that is specified for every equilibrium state of a substance. Like energy, entropy is an abstract concept that is extensively used in thermodynamic analysis. Entropy represents the microscopic disorder or uncertainty of a system. Because entropy is a property, the change in entropy in going from one state to another is the same for all processes. The SI unit for specific entropy is J/K.

The reciprocating engines are considered first in the following section to be followed by a discussion on gas turbines.

10.2.1 Reciprocating Engines

The two types of reciprocating ICEs are the spark-ignition engine (SI) and the compression-ignition (CI) engine. The two engines are commonly known as gasoline/petrol engine and diesel engine, based on the type of fuel used for combustion. The difference in the two engines is in the method of initiating the combustion and

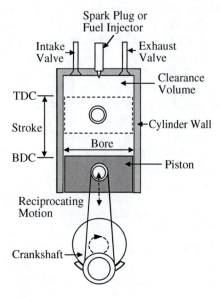

FIGURE 10.4 A reciprocating IC engine.

in the processes of the cycle. In a SI engine, a mixture of air and fuel is drawn in, and a spark plug ignites the charge. The intake of the engine is called the *charge*. In a CI engine, air is drawn in and compressed to such a high pressure and temperature that combustion starts spontaneously when, later, fuel is injected. The SI engines are relatively light and lower in cost and are used for lower-power engines, as in conventional automobiles. The CI engines are more suitable for power conversion in higher power range, such as in trucks, buses, locomotives, ships, and in auxiliary power units. Their fuel economy is better than the SI engines, justifying their use in higher-power applications.[1]

A sketch of a representative reciprocating ICE, including the special terms standard for such engines, is given in Figure 10.4. The engine consists of a piston that undergoes a reciprocating motion within the engine cylinder. The position of the piston at the bottom of the cylinder when the volume inside is maximum is known as the *bottom dead center (BDC)*. The position of the piston at the top of the cylinder when the volume inside is minimum is called the *top dead center (TDC)*. This cylinder minimum volume when the piston is at TDC is known as the *clearance volume*. A crank mechanism converts the linear motion of the piston into rotary motion and delivers the power to the crankshaft. The volume swept by the piston as it moves from the TDC to the BDC is known as the *displacement volume*, which is a parameter commonly used to specify the size of an engine. The *compression ratio* is defined as the ratio of the volume at BDC to the volume at TDC.

The diameter of the cylinder is called the *bore*. The bore in automotive SI engines is typically between 70 to 100 mm. Too small a bore leaves no room for valves, while an exceedingly large bore means more mass and longer flame travel time. The smaller bores enable higher rpm of the engines. The vertical distance traversed by the piston from the BDC to the TDC is called the *stroke*. The stroke is typically

TABLE 10.1
Arrangement of Automotive Engine Cylinders

Number of Cylinders	Cylinder Arrangement
3	Inline
4	Inline, flat
5	Inline
6	Inline, flat, V (narrow, 60°, 90°)
8	V (90°)
10	V (90°)
12	V, flat (for exotics)

between 70 to 100 mm. Too short a stroke means that there will not be enough torque. The length of the stroke is limited by piston velocities. The minimum displacement of a cylinder can be 250 cc, while the maximum can be up to 1000 cc. The acceptable bore and stroke lengths lead to multiple cylinder engines. The multiple cylinders can be arranged inline, flat, or in a V-shaped configuration, depending on the number of cylinders. The typical arrangements are given in Table 10.1. For a good primary rotational balance, the power strokes of the multiple cylinders are equally spaced. The engine arrangements that have good primary balance are Inline 4 and 6 cylinders, 90° V 8 cylinders, and flat 4 and 6 cylinders. The arrangements that have poor primary balance are 90° V 6 cylinders and Inline 3 cylinders. In the arrangements with poor primary balance, counterrotating balance shafts are used to cancel vibration.

The valve arrangement in the cylinder is known as the valve train. The valve train can be overhead valve (OHV), single overhead cam valve (SOHC), or dual overhead cam valve (DOHC). The OHV has cam, push rod, rocker, and valve; the SOHC has cam, rocker, and valve; while the DOHC has cam, rocker, valve, cam, and valve. There can be 2, 3, 4, or 5 valves in a cylinder. The number of valves selected depends on the trade-off between flow and complexity.

10.2.1.1 Practical and Air-Standard Cycles

Automobile ICEs are typically four-stroke engines, where the piston executes four strokes of the cylinder for every two revolutions of the crankshaft. The four strokes are *intake, compression, expansion* or *power*, and *exhaust*. The operations within the four strokes are illustrated in the pressure volume diagram of Figure 10.5. Numbers one through five in the diagram represent the distinct state points between processes of the cycle. The *intake* is the process of drawing the charge into the cylinder with the intake valve open. The working fluid is compressed in the *compression* stage, with the piston traveling from the BDC to the TDC. Work is done by the piston in the compression stage. In the next stage, heat is added during ignition of the compressed fluid, with the respective processes of ignition for SI and CI

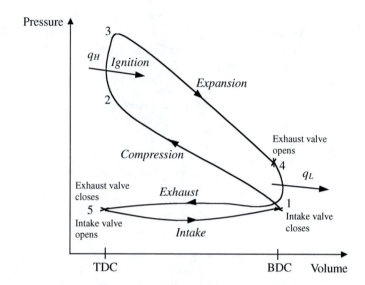

FIGURE 10.5 Pressure-volume diagram of a reciprocating-type IC engine.

engines. The next stage is the *expansion* process, which is also known as the power stroke. In this stroke, work is done by the charge. The *exhaust* process starts at the BDC with the opening of the exhaust valve. Heat is rejected from the engine during the exhaust process.

Practical cycles entail significant complexity due to the irreversibilities associated with friction, pressure, and temperature gradients, heat transfer between the gases and the cylinder walls, and work required to compress the charge and exhaust the products of combustion. The complexity of the process typically calls for computer simulation for a performance analysis. However, significant insight can be gained into the processes by making simplifying assumptions about the behavior of the processes that make up the cycle. Idealized processes can substitute the combustion and expansion processes within the cylinder. These idealized cycles are known as air-standard cycles. The air-standard analysis assumes that the working fluid is an ideal gas, the processes are reversible, and the combustion and exhaust processes are replaced by a heat transfer with an external source. A brief description of the two air-standard cycles, the Otto cycle and the Diesel cycle, are given in the following.

10.2.1.2 Air-Standard Otto Cycle

The Otto cycle is the idealized air-standard version of the practical cycle used in SI engines. The air-standard Otto cycle assumes that heat addition occurs instantaneously under constant volume when the piston is at the TDC. The cycle is illustrated on a *p-v* (pressure-volume) diagram in Figure 10.6. The *intake stroke* starts with the intake valve opening at the TDC to draw a fresh charge into the cylinder. The intake valve is open between 1 to 5 to take the fresh charge, which is a mixture of fuel and air. The volume of the cylinder increases as the piston moves down to allow more charge into the cylinder. The stroke ends with the piston reaching the BDC

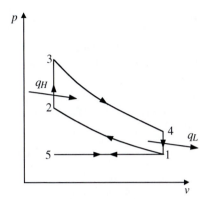

FIGURE 10.6 *p-v* diagram of an air-standard Otto cycle.

when the intake valve closes at that position. This state point at the BDC is labeled as 1. In the next process between 1 and 2, work is done on the charge by the piston to compress the charge, thereby increasing its temperature and pressure. This is the *compression cycle*, when the piston moves up with both valves closed. Process 1 to 2 is an isentropic (constant entropy) compression, as the piston moves from the BDC to the TDC. The combustion starts near the end of the compression stroke in SI engines, when the high-pressure, high-temperature fluid is ignited by the spark plug. The pressure thus rises at constant volume to state point 3. Process 2 to 3 is the rapid combustion process when heat is transferred at constant volume to the air from the external source. The next stroke is the *expansion* or *power stroke*, when the gas mixture expands, and work is done by the charge on the piston, forcing it to return to the BDC. Process 3 to 4 represents the isentropic expansion when work is done on the piston. The final stroke is the *exhaust stroke*, which starts with the opening of the exhaust valve near 4. During Process 4 to 1, the heat is rejected, while the piston is at BDC. Process 1 to 5 represents the exhaust of the burnt fuel at essentially constant pressure. At 5, the exhaust valve closes, and the intake valve opens; the cylinder is now ready to draw in fresh charge for a repeat of the cycle.

SI engines can be four-stroke or two-stroke engines. Two-stroke engines run on the two-stroke Otto cycle, where the intake, compression, expansion, and exhaust operations are accomplished in one revolution of the crankshaft. The two-stroke cycles are used in smaller engines, such as those used in motorbikes.

Most SI engines, or gasoline engines as more commonly known, run on a modified Otto cycle. The air–fuel ratio used in these engines is between 10/1 to 13/1. The compression ratios are in the range of 9 to 12 for most production vehicles. The compression ratio of the engine is limited by the octane rating of the fuel. If the octane number of the fuel is too low, a high compression ratio may lead to auto-ignition of the air–fuel mixture during compression, which is completely undesirable in a SI engine. SI engines were originally developed by limiting the amount of air allowed into the engine using a carburetor. The carburetor is the throttling valve placed on the air intake. However, fuel injection, which is used for diesel engines, is now common for gasoline engines with SI. The control issue for fuel injection systems is to compute the mass flow rate of air into the engine at any instant of time and to mix the correct amount of gasoline with it, such that the air and fuel mixture is right for the engine running condition. In recent years, requirements to meet the strict exhaust gas emission regulations have increased the demand for fuel injection systems.

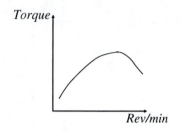

FIGURE 10.7 Torque–speed characteristics of a gasoline engine.

The torque–speed characteristics of a SI or gasoline engine are shown in Figure 10.7. The engine has a narrow high torque range, which also requires high enough rpm of the engine. The narrow high-torque region burdens the transmission gear requirements of SI engines.

SI engines are widely used in automobiles, and continuous development has resulted in engines that easily meet current emission and fuel economy standards. Currently, SI engines are of the lowest cost engines, but the question remains as to whether it will be possible to meet future emission and fuel economy standards at a reasonable cost. The SI engine also has a few other drawbacks, which include the throttling plate used to restrict the air intake. The partial throttle operation is poor in SI engines due to throttle irreversibility, a problem that is nonexistent in diesel engines. In general, the throttling process leads to a reduction in efficiency of the SI engine. The losses through bearing friction and sliding friction further reduce the efficiency of the engine.

10.2.1.3 Air-Standard Diesel Cycle

The practical cycles in diesel engines are based on the Diesel cycle. The air-standard Diesel cycle assumes that heat addition takes place at constant pressure, while heat rejection occurs under constant volume. The cycle is shown on a *p-v* (pressure-volume) diagram in Figure 10.8.

The cycle begins with the *intake* of fresh air into the cylinder between 5 to 1. The intake valve is open between 5 and 1. The next process, 1 to 2, is the same as in the Otto cycle, when isentropic (constant entropy) *compression* takes place as the piston moves from the BDC to the TDC. With a sufficiently high compression ratio, the temperature and pressure of air reaches such a level that the combustion starts spontaneously due to injection of fuel near the end of the compression stroke. The heat is transferred to the working fluid under constant pressure during combustion in Process 2 to 3, which also makes up the first part of the *expansion* or power stroke. The isentropic expansion in Process 3 to 4

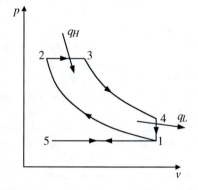

FIGURE 10.8 *p-v* diagram of an air-standard Diesel cycle.

makes up for the rest of the power stroke. The *exhaust* valve opens at state point 4, allowing the pressure to drop under constant volume during Process 4 to 1. Heat is rejected during this process, while the piston is at BDC. The exhaust of the burnt fuel takes place during Process 1 to 5 at essentially constant pressure. The exhaust valve then closes, the intake valve opens, and the cylinder is ready to draw in fresh air for a repeat of the cycle.

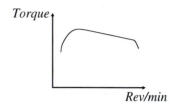

FIGURE 10.9 Torque–speed characteristics of a diesel engine.

The nominal range of compression ratios in CI engines is 13/1 to 17/1, and the air–fuel ratios used lie between 20/1 and 25/1. The higher compression ratio aided by the work produced during combustion results in higher efficiency in diesel engines compared to gasoline engines. Efficiencies in diesel engines can be as high as 40%. The diesel engine has a lower specific power than the gasoline engine. Diesel engines also have a broad torque range, as shown in Figure 10.9.

The major drawbacks of diesel engines include the requirement of stronger and heavier components that increase the mass of the engine and the speed limitation of the injection and flame propagation time. The improvements in diesel engines are directed toward reducing nitrogen oxides in the exhaust; and reducing the noise, vibration, and smell of the engine. Recent automotive diesel engines developed addressing the aforementioned issues made them excellent candidates for HEV applications.

10.2.1.4 Example IC Engines in HEVs

Toyota Prius
 1.5 liter DOHC 16 valve Inline 4 cylinders
 Bore × Stroke — 75.0 × 84.7
 Variable valve timing
 75 hp @ 4500 rpm
 82 ft-lb @ 4200 rpm
 13:1 compression
 Runs on Atkinson cycle with extended expansion cycle and reduced compression cycle; the cycle improves thermal efficiency at the expense of specific output
 EPA gas mileage — 52 (highway)/45 (city) (for the ICE only)
 Gross weight — 2775 lb

Honda Insight
 1.0 liter SOHC 12 valve Inline 3
 Bore × Stroke — 72.0 × 81.5
 Variable valve timing and lift control (VTEC) — two sets of cam lobes
 67 hp @ 5700 rpm
 66 ft-lb @ 4800 rpm
 10.8:1 compression ratio

Stratefied charge: 14.7:1 to 22:1 (richer near spark plug)
Lean burn at lower rpm for more torque and efficiency
EPA gas mileage —70 (highway)/61 (city) (for the ICE only)
Gross weight — 1940 lb

The above gasoline engines used in HEVs can be compared with a diesel engine used in a conventional vehicle, specifics of which are given in the following.

VW Jetta TD (Turbo Diesel)
1.9 liter SOHC 8 valve Inline 4, turbocharged
19.5:1 compression
Bore × Stroke — 79.5 × 95.5
90 hp @ 3750 rpm
155 ft-lb @ 1900 rpm
EPA gas mileage — 49 (highway)/41 (city)

The excellent engine characteristics of the VW Jetta ICEV show that diesel engines can be good candidates for thermal engines in hybrid vehicles.

10.2.2 GAS TURBINE ENGINE

Gas turbines are used for stationary power generation as well as for transportation applications, such as in aircraft propulsion and marine power plants. Gas turbines can operate on either an open- or closed-cycle basis. In the open-cycle type, the working fluids gain their energy from combustion of fuel within the engine, whereas in a closed type, the energy input by heat transfer is from an external source. The open type is used in vehicle propulsion systems and will be considered here. The use of gas turbines in transportation applications can be attributed to the favorable power output-to-weight ratio of gas turbines, which again makes it a viable candidate for HEVs.

The gas turbine engine runs on a Brayton cycle, which uses constant pressure heat transfer processes with isentropic compression and expansion processes in between. The major components of a gas turbine are shown in Figure 10.10a, along with the directions of energy flow. The corresponding p-v diagram for an air-standard cycle is illustrated in Figure 10.10b, which ignores the irreversibilites in air circulation through the components of the gas turbine. The constant entropy is represented by $s = c$.

The working fluid of air is considered to be an ideal gas in the air-standard analysis. At the beginning of the cycle at State 1, atmospheric air is continuously drawn in and compressed in the compressor to raise its pressure and temperature. The compressor is usually of the radial flow type for automotive applications. The air then moves into the combustion chamber or combustor, where combustion takes place with fuel injected to raise the temperature of the air. The high temperature–high pressure mixture is then expanded and cooled in the turbine, which produces power and delivers work. The hot gas turbine exhaust gas is utilized in a recuperator to

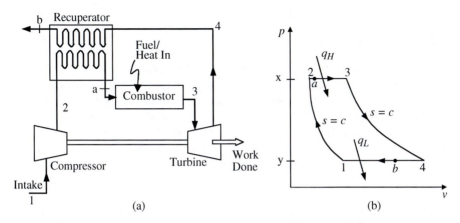

FIGURE 10.10 (a) Air-standard gas turbine cycle. (b) *p-v* diagram of ideal Brayton cycle.

preheat the air exiting the compressor before entering the combustor. This reduces the fuel needed in the combustor and increases the overall efficiency of the system. In the open-type gas turbine, fresh air is drawn in and exhaust gases are purged after going through the recuperator in each cycle. The turbine is designed to deliver work output greater than the required compressor work input. The excess work available at the shaft is used to move a vehicle in automotive gas turbines or to generate electricity in stationary applications. The enclosed area in the figure represents the net work output. In particular, the area 12xy1 in the *p-v* diagram represents the compressor work input per unit mass and the area 34yx3 represents the turbine work output per unit mass.

The power output of a turbine is controlled through the amount of fuel injected into the combustor. Many turbines have adjustable vanes and gearing to decrease fuel consumption during partial load conditions and to improve acceleration. The major advantage of a gas turbine is that the only moving part is the rotor of the turbine. The turbine has no reciprocating motion, and consequently, runs smoother than a reciprocating engine. Another advantage of gas turbines is multifuel capability. The turbine has the flexibility of burning any combustible fuel injected into the airstream, because continuous combustion is not heavily reliant on combustion characteristics of the fuel. The fuel in a gas turbine burns completely and cleanly, which keeps emissions at a low level.

The gas turbine engine has a few drawbacks, which have prevented its wide-spread use in automotive applications. The complicated design of gas turbines increases the manufacturing costs. The response time of the gas turbine to changes in throttle request is slow, relative to a reciprocating engine. The efficiency of the gas turbine decreases at partial throttle conditions, making it less suitable for low-power applications. The turbine requires intercoolers, regenerators, and reheaters to reach efficiencies comparable to SI or CI engines, which adds significant cost and complexity to a gas turbine engine.

10.3 DESIGN OF AN HEV

10.3.1 HYBRID DRIVETRAINS

The drivetrain architecture and control technique for an HEV depends on the desired requirements, including, but not limited to, performance, range, and emission. The performance requirements of initial acceleration, cruising velocity, maximum velocity, and gradability dictate the design of power and energy requirements of the engine and motor. The energy and power requirements can also be specified in terms of multiple driving schedules that have the worst-case demands embedded in them. The energy required by the drivetrain to meet the range specification dictates the choice of the energy source unit, which can be a battery pack or a combination of battery and ultracapacitors. Meeting the emission standards depends solely on heat engine emission characteristics, because the electric motor has zero emission.

The power required in an HEV comes from a combination of the electric motor and heat engine outputs. The mission of the vehicle plays a big role in apportioning the power requirement between the electric motor and heat engine. An HEV designed for urban commute transportation will have a different combination of drivetrain subsystems than a family sedan designed for urban and highway travel. The power requirement of the electric motor and the heat engine for an urban commute vehicle will definitely be lower than that required for cars designed for highway travel. For urban vehicles used for daily commutes of less than 100 miles, the designer must also question the advantage of a hybrid vehicle compared to a battery-powered EV. The design engineer evaluates the design trade-offs based on the mission and specifications, and selects one of the series or parallel configurations discussed earlier in Section 10.1. The subsystems of the drivetrain have individual control units, and the components are coordinated through a supervisory control unit. The mission of the vehicle also dictates the type of control to be employed for the vehicle. For example, highway vehicles require a control strategy designed to extend the range of the vehicle. The designed system must be capable of handling all real-world situations within the limits of the design requirements. Appropriate safety measures must also be incorporated in the controller to handle situations when vehicle capabilities exceed the roadway condition.

10.3.2 SIZING OF COMPONENTS

The sizing of the components of the electrical system and the mechanical system starts once the drivetrain architecture is laid out based on the mission of the vehicle. In a series HEV, the electrical system design is the same as that of an EV. The IC engine size is specified for keeping the batteries charged. The sizing of the components of a parallel HEV is much more complex. If the vehicle is designed with heavier biasing on the IC engine, then the batteries can be downsized and reconfigured for maximum specific power instead of maximum specific energy. The battery and the motor serve to supply peak power demands during acceleration and overtaking without being discharged. The battery also acts as a reservoir for the regenerative braking energy. Ultracapacitors can be used instead of batteries, provided they meet the requirements during peak power demands. If the vehicle is to be more battery biased, then the system is configured such that the batteries will reach about 80% DoD at the end of the longest trip. Once the power requirements of the electrical and mechanical systems are apportioned for the parallel HEV, the electrical components are designed based on the power designated for the electrical system using the same design philosophy as that used for the EV components. A philosophy parallel to that used for EV can be used to design mechanical subsystems, where the components are sized based on initial acceleration, rated cruising velocity, maximum velocity, and maximum gradability.[2] The gear ratio between the IC engine and the wheel shaft of a parallel HEV can be obtained by matching the maximum speed of the IC engine to the maximum speed of the driveshaft. A single gear transmission is desired to minimize complexity. The sizing of the components of a parallel HEV is discussed below.

10.3.2.1   Rated Vehicle Velocity

In an HEV, the electric motor primarily serves to meet the acceleration requirement, while the IC engine delivers the power for cruising at rated velocity, assuming that the battery energy is not sufficient to provide the required power throughout the desired range. Therefore, the IC engine size is determined by the vehicle cruising power requirement at its rated velocity, independent of the electric motor power capacity. Thus, the IC engine size should be determined first in the case of HEV, and its size can be used to reduce the power requirement of the electric motor responsible for vehicle acceleration characteristics.

The road load characteristics developed in Chapter 2 and the force-velocity characteristics of the IC engine (derived from the torque–speed characteristics such as shown in Figure 10.7 or 10.9) are useful in sizing the IC engine. Figure 10.11 shows example curves of IC engine characteristics with engine displacement as a parameter, along with the road load characteristics for an assumed grade and vehicle parameters. The correct IC engine size is determined from the intersection of the worst-case road load characteristics with the IC engine force-velocity profile at rated velocity, plus allowing a nominal 10% margin for battery pack recharging.[2] The exact amount of margin needed is the subject of a more complicated analysis, involving vehicle driving cycles, battery capacity, battery charge and discharge characteristics, and generator characteristics.

10.3.2.2   Initial Acceleration

The electric motor with its higher peak power capabilities is more heavily used during initial acceleration. The mechanical power available from the IC engine can be blended with the electric motor power for acceleration, thereby reducing the power requirement of the electric motor. The power required from the motor depends on the velocity at which torque blending from the two propulsion units starts. Figure 10.12 shows the effect of torque blending on the electric motor rated power requirement during initial acceleration with a single-gear transmission for the IC engine. The figure shows that there will be little power contribution from the engine until a minimum critical velocity v_{cr} mi/h of the vehicle is reached due to its poor

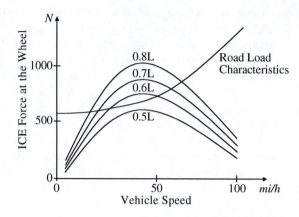

FIGURE 10.11 Typical IC engine force-velocity characteristics and road load characteristics.

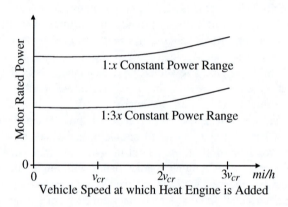

FIGURE 10.12 Electric power requirement as a function of vehicle speed at which IC engine is added.

low-speed torque capability. Therefore, torque blending should start after the vehicle has attained the critical velocity with single-gear transmission, avoiding the use of the engine for initial acceleration as much as possible without significantly increasing the rating of the electric motor. The power requirement from the electric motor increases nonlinearly with speed if IC engine torque blending is delayed beyond v_{cr} mi/h. The qualitative figure shows two curves, with the extent of the constant power region as a parameter, x being an integer number. The lower curve has a much wider constant power region than the upper curve. The curves emphasize the need for an extended constant power region of operation of the electric motor to minimize its size. The power requirement of the IC engine determined from rated vehicle velocity conditions would typically be enough to provide the initial acceleration in combination with the electric motor.

10.3.2.3 Maximum Velocity

The power requirement from the propulsion system at maximum velocity is $F_{TR} \cdot v_{max}$, which is supplied by a combination of the engine and the electric motor. The power requirement is given by Equation 9.8, repeated here for convenience.

$$P_{TR,max} = F_{TR} \cdot v_{max} = mgv_{max} \sin\beta + \left[mgC_1 + \frac{\rho}{2} A_F C_D \right] v_{max}^3 + mgv_{max}C_0$$

The electric motor power required to meet maximum velocity conditions can be uniquely de ned by subtracting the engine po wer determined for cruising at rated velocity from the maximum velocity power requirement. The electric motor power requirement calculated in this step would, in general, be less than the power requirement for the initial acceleration, unless the maximum velocity requirements are stringent, such as high speed on a steep grade.

10.3.2.4 Maximum Gradability

The maximum gradability condition must be checked once the sizing of the IC engine and the electric motor is done from the previous three requirements. The maximum gradability of a vehicle is given by

$$\text{Max. \% grade} = \frac{100 \ F_{TR}}{\sqrt{(mg)^2 - F_{TR}^2}}$$

If the condition is not met, then the size of the engine or the motor or both must be increased or the gear ratio must be changed to meet the gradability requirements.

Although the design philosophy outlined above states that the IC engine sizing primarily comes from the rated cruising velocity, and the electric motor sizing comes from the initial acceleration, the practical design involves extensive computer simulation using various drive cycles, parameters of the vehicle, and characteristics of chosen battery, motor, generator, and IC engine. As in all systems cases, the sizing and design of the components of the EV and the HEV is an iterative process that ends when all the design requirements are met. However, the discussions presented on design provide the theoretical basis for initial estimates, avoiding unnecessary oversizing of components.

The speci cations of the components of an e xample HEV (Toyota Prius) are given in the following. The example is not necessarily for the best or optimum HEV, but merely a representative one from today's production HEVs. The Toyota Prius uses a series–parallel combination hybrid, as shown in Figure 10.3. During starting and moving under light load, the electric motor supplies the power, while the IC engine is turned off. The IC engine also turns off during deceleration or regenerative braking, saving energy and reducing pollution. During normal operation, the IC engine provides the power with assistance from the electric motor. Part of the power generated by the IC engine is used to generate electricity through the series generator, which supplies power to the motor (refer to Figure 10.3 to check the power flow direction). The generated power is also used to keep the batteries charged. During full throttle operation, extra energy is drawn from the battery to overcome demanding situations. The Toyota Prius performs like a regular ICEV, but the addition of the electric motor power unit helps reduce the CO_2, HC, CO, and NO_x emissions. The battery capacity of the vehicle is rather low to give a good range by drawing energy only from the batteries.

Specifications of Toyota Prius

Performance:
0–400 m in 19.4 s
Top speed — 160 km/h

Dimensions:
Overall height, width, and length — 57.6 in × 66.7 in × 169.6 in
Aerodynamic drag coefficient — $C_D = 0.29$
Curb weight — 1258.5 kg

IC engine specifications:
1500 cc four-stroke engine using high expansion Atkinson cycle
DOHC 16 valve, variable valve timing, electronic fuel injection
Compression ratio — 13:1
Peak power — 42.5 kW at 4000 rev/m
Peak torque — 102 Nm at 4000 rev/m

Electric motor specifications:
PM synchronous motor
Maximum power — 30 kW
Rated speed — 940 rev/m
Constant power range — 940–2000 rev/m
Peak torque — 305 Nm from standstill to 940 rev/m
System voltage — 288 V

Energy storage:
Battery pack — 40 NiMH batteries
Battery output voltage — 273.6 V (228, 1.2 V cells)
Battery energy capacity — 1.9 kWh
Rated battery capacity — 6.5 Ah

REFERENCES

1. Moran, M.J. and Shapiro, H.N., *Fundamentals of Engineering Thermodynamcis*, 3rd ed., John Wiley & Sons, New York, 1995.
2. Howell, J.R. and Buckius, R.O., *Fundamentals of Engineering Thermodynamics*, 2nd ed., McGraw-Hill, New York, 1992.
3. Ehsani, M., Rahman, K.M., and Toliyat, H.A., Propulsion system design for electric and hybrid vehicles, *IEEE Transactions on Industrial Electronics*, Vol. 44, No. 1, February, 1997, pp. 19–27.

PROBLEM

10.1

An HEV has the following parameter values: $\rho = 1.16$ kg/m³, m = 692 kg, $C_D = 0.2$, $A_F = 2$ m², $g = 9.81$ m/s², $C_0 = 0.009$, and $C_1 = 1.75*10^{-6}$ s²/m². The type of IC engine that will be used for the vehicle has the force (at wheel) vs. velocity characteristics of $F_{TR} = 2.0 \sin 0.0285x$ N for $5 < x < 100$, where x is the vehicle speed in mi/h. Determine the displacement of the ICE for a rated cruising velocity of 60 mi/h on a 2% slope.

Index